THE VOICE
OF THE FUTURE . . .

THE UNIVERSE MUST BE FULL OF VOICES, calling from star to star in a myriad tongues. One day we shall join that cosmic conversation, though it may be ages before we have much to contribute to it.

There may even be seas on the Moon; if there are, they will be of dust. Dry and slippery as talcum, it could have gathered during the ages in low-lying areas, where it may be waiting to trap future explorers. You could walk across it with the aid of skis or snow shoes, and one day there may be paddle-wheel moon boats sailing the lunar seas.

Somewhere on Mars there will be a street or a city park where it's Friday on one side and Saturday on the other. It will be quite a tourist attraction but a great nuisance to people who have to live with it. The fact that this doesn't happen on Earth is pure luck. It must occur on *all* inhabited planets that do not possess oceans in which unwanted days can be drowned.

ARTHUR C. CLARKE
VOICES FROM THE SKY

PUBLISHED BY POCKET BOOKS NEW YORK

ACKNOWLEDGMENTS

"Space Flight and the Spirit of Man," reprinted by courtesy of *Astronautics,* a publication of the American Rocket Society, from Space Flight Report to the Nation Issue, October 1961.

"The World of the Communications Satellite," courtesy of *Astronautics & Aeronautics,* a publication of the American Institute of Aeronautics and Astronautics, February 1964.

"Telstar Over Broadway." Reprinted with permission of and copyright by *Playbill,* America's magazine for theatergoers.

"Ships for the Stars." Reprinted from *The Rotarian* magazine for May 1962.

"Seas of Tomorrow," reprinted with the permission of Gilberton Company, Inc., New York.

"How I Lost a Billion Dollars in My Spare Time," copyright 1962 by *Rogue Magazine.* Reprinted by permission.

"The Social Consequences of the Communications Satellites," "The Light of Common Day." Reprinted by permission of *Horizon* magazine.

"H. G. Wells and Science Fiction." Copyright © 1962 by Washington Square Press, Inc. Introduction to *The Invisible Man* and *War of the Worlds.*

"Extraterrestrial Relays." First published in *Wireless World* October 1945. Reprinted by permission.

"Class of '00." Reprinted from 3M Company *Tartan* by special permission.

"The Electronic Revolution." © 1962 by The New York Times Company. Reprinted by permission.

"The Meddlers" and "Beyond Centaurus" originally appeared in *Playboy Magazine.*

"Memoirs of an Armchair Astronaut" and "The Playing Fields of Space" originally appeared in *Holiday.*

POCKET BOOKS, a Simon & Schuster division of
GULF & WESTERN CORPORATION
1230 Avenue of the Americas, New York, N.Y. 10020

Copyright 1945, © 1961, 1962, 1964, 1965 by Arthur C. Clarke

Published by arrangement with Harper & Row, Publishers, Inc.
Library of Congress Catalog Card Number: 65-20992

ISBN: 0-671-82141-5

First Pocket Books printing February, 1980

10 9 8 7 6 5 4 3 2 1

POCKET and colophon are trademarks of Simon & Schuster.

Printed in the U.S.A.

TO GEORGE
best and most patient of editors

Contents

Preface 1

The essays in this section are all concerned with astronautics and astronomy.

The first was commissioned by the editor of the American Rocket Society's magazine Astronautics for a special issue published in connection with the society's Space Flight Report to the nation at the New York Coliseum, October 1961. It was later reprinted in Reader's Digest.

"The Uses of the Moon" was written in May 1961, and it took the editors of Harper's magazine an unconscionable amount of time before they decided to publish it. They probably felt that their more conservative readers would regard lunar colonization as a joke—and they were right. (For one reaction, see "Dear Sir—".) Several indignant letters were fowarded to me, and others were printed in the correspondence column. It is therefore with considerable satisfaction that I can record the article's choice as the runner-up for the best magazine science writing of the year in the Westinghouse–American Academy of Science Annual Award.

Events since 1961 have not dated the conclusions; indeed, the Apollo Project has made them even more timely.

Space Flight
and the Spirit of Man

It is exactly fifteen years since, at the October 1946 meeting of the British Interplanetary Society, I presented the first version of my paper *The Challenge of the Spaceship,* an inquiry into the cultural and philosophical implications of astronautics.[1] At the time, as the title indicates, I was somewhat under the influence of Professor Toynbee, having just attended a lecture he had given at the Senate House, University of London, on "The Unification of the World." He had opened my eyes to the highly parochial view we Westerners take of human history, which is best summed up by our attitude that *we* discovered the rest of the world. Above all, however, I was struck by Toynbee's emphasis on "challenge and response" as shaping the rise and fall of civilizations, and it seemed to me that we would be presented with a classic example of this when the Space Age opened. Here without question was the greatest physical challenge that life on this planet had faced since the distant days when it emerged from the sea and invaded that other hostile environment, the arid, sun-scorched land.

As I went on to consider the possibilities opened up by this new field of exploration, my mind was inevitably drawn to the great voyages of discovery of the fifteenth and sixteenth centuries. These were not only voyages of discovery, but of escape; they liberated men's minds

[1] The latest version is the opening essay in the book of the same name. (Harper & Brothers, 1959.)

11

from the long trance of the Middle Ages, and fueled the fires of the Renaissance. Perhaps something similar would happen with space flight; looking toward a future which, in 1946, still seemed very distant, I wrote the following words:

> With the expansion of the world's mental horizons may come one of the greatest outbursts of creative activity ever known. The parallel with the Renaissance, with its great flowering of the arts and sciences, is very suggestive. "In human records," wrote the anthropologist J. D. Unwin, "there is no trace of any display of productive energy that has not been preceded by a display of expansive energy. Although the two kinds of energy must be carefully distinguished, in the past they have been united in the sense that one has developed out of the other." Unwin continues with this quotation from Sir James Frazer: "Intellectual progress, which reveals itself in the growth of art and science . . . receives an immense impetus from conquest and empire." Interplanetary travel is now the only form of "conquest and empire" compatible with civilization. Without it, the human mind, compelled to circle forever in its planetary goldfish bowl, must eventually stagnate.

Now that we are well into the space age, and achievements which in 1946 seemed to belong to the remote future are milestones in the past. it is time to ask if these predictions of a cultural revival can still be justified—and even if they already show signs of coming true.

That the world is now space conscious, to an extent which would have seemed unbelievable only a few years ago, is a statement that needs no proof. But it is not yet space *minded*. By this, I mean that the general public still thinks of space activities almost exclusively in terms of military strength and international prestige. These matters are, of course, vitally important; yet in the long run, if there is a long run, they will be merely the ephemeral concerns of our neurotic age. In the sane society which we have to build if we are to survive, we must forget spacemanship and concentrate on space.

Unfortunately, altogether too many educators, intellectuals and other molders of public opinion, still regard space as a terrifying vacuum, instead of a frontier with infinite possibilities. Typical of this attitude, though seldom so clearly expressed, is the following passage from Professor Lewis Mumford's *The Transformation of Man:*

> Post-historic man's starvation of life would reach its culminating point in interplanetary travel. . . . Under such conditions, life would again narrow down to the physiological functions of breathing, eating and excretion. . . . By comparison, the Egyptian cult of the dead was overflowing with vitality; from a mummy in his tomb one can still gather more of the attributes of a full human being than from a spaceman.

The almost laughable falsehood of this passage was demonstrated by Commander Shepard's famous exclamation "What a beautiful sight!" as his Mercury capsule arced over the Caribbean. I would maintain that these words are enough to settle the matter, but it must be admitted that most people would prefer more substantial evidence for the benefits of manned space flight.

Let me first dispose of one argument for man in space that is frequently put forward, and which only confuses the issue. It is often suggested that the complexity and unreliability of automatic space probes will make it impossible to dispense with human astronauts, even if they merely serve as trouble shooters. This is a short-sighted view; in the not-too-distant future—perhaps only fifty years from now—we will have robots as good as any flesh-and-blood explorers. The frequent and predictable failures of the next decade's automatic astronauts must not blind us to the fact that they will be only clumsy, moronic toys compared with their successors half a century hence. The justification of man in space must depend not upon the deficiencies of his machines, but upon the positive advantages that he, personally, will gain from going there.

There is no point in exploring—still less colonizing—a hostile and dangerous environment unless it opens up new opportunities for experience and spiritual enrichment. Mere survival is not sufficient; there are already enough examples on this planet of societies that have been beaten down to subsistence level by the forces of nature. The questions which all protagonists of space flight have to ask themselves, and answer to their own satisfaction, are these: What can the other planets offer that we cannot find here on Earth? Can we do better, on Mars or Venus, than the Eskimos have done in the Arctic? And the Eskimos, it is worth reminding ourselves, have done very well indeed; a dispassionate observer might reasonably decide that they are the only really civilized people on this planet.

The possible advantages of space can be best appreciated if we turn our backs upon it and return, in imagination, to the sea. Here is the perfect environment for life—the place where it originally evolved. In the sea, an all-pervading fluid medium carries oxygen and food to every organism; it need never hunt for either. The same medium neutralizes gravity, insures against temperature extremes, and prevents damage by too-intense solar radiation—which must have been lethal at the Earth's surface before the ozone layer was formed.

When we consider these facts, it seems incredible that life ever left the sea, for in some ways the dry land is almost as dangerous as space. Because we are accustomed to it, we forget the price we have had to pay in our daily battle against gravity. We seldom stop to think that we are still creatures of the sea, able to leave it only because, from birth to death, we wear the water-filled space suits of our skins.

Yet until life had invaded and conquered the land, it was trapped in an evolutionary cul-de-sac—for intelligence cannot arise in the sea. The relative opacity of water, and its resistance to movement, were perhaps the chief factors limiting the mental progress of marine creatures. They had little incentive to develop keen vision (the most subtle of the senses, and the only long range one) or manual dexterity. It will be most interesting to see if there are any exceptions to this, elsewhere in the universe.

Even if these obstacles do not prevent a low order of intelligence arising in the sea, the road to further development is blocked by an impassable barrier. The difference between man and animals lies not in the possession of tools, *but in the possession of fire.* A marine culture could never escape from the Stone Age and discover the use of metals; indeed, almost all branches of science and technology would be forever barred to it.

Perhaps we would have been happier had we remained in the sea (the porpoises seem glad enough to have returned, after sampling the delights of the dry land for a few million years) but I do not think that even the most cynical philosopher has ever suggested that we took the wrong road. The world beneath the waves is beautiful, but it is hopelessly limited, and the creatures who live there are crippled irremediably in mind and spirit. No fish can see the stars; but we will never be content until we have reached them.

There is one point, and a very important one, at which the evolutionary parallel breaks down. Life adapted itself to the land by unconscious, biological means, whereas the adaptation to space is conscious and deliberate, made not through biological but through engineering techniques of infinitely greater flexibility and power. At least, we think it is conscious and deliberate, but it is often hard to avoid the feeling that we are in the grip of some mysterious force or *Zeitgeist* that is driving us out to the planets, whether we wish to go or not.

Though the analogy is obvious, it cannot be *proved,* at this moment of time, that expansion into space will produce a quantum jump in our development as great as that which took place when our ancestors left the sea. From the nature of things, we cannot predict the new forces, powers, and discoveries that will be disclosed to us when we reach the other planets or can set up laboratories in space. They are as much beyond our vision today as fire or electricity would be beyond the imagination of a fish.

Yet no one can doubt that the increasing flow of knowledge and sense impressions, and the wholly new types of experience and emotion, that will result from space travel will have a profoundly stimulating effect upon the human psyche. I have already referred to our age as a neurotic one; the "sick" jokes, the decadence of art forms, the flood of anxious self-improvement books, the etiolated cadavers posing in the fashion magazines—these are minor symptoms of a malaise that has gripped at least the Western world, where it sometimes seems that we have reached *fin de siècle* fifty years ahead of the calendar.

The opening of the space frontier will change all that, as the opening of any new frontier must do. It has saved us, perhaps in the nick of time, by providing an outlet for dangerously stifled energies. In William James's famous phrase, it is the perfect "moral equivalent of war."

From time to time, alarm has been expressed at the danger of "sensory deprivation" in space. Astronauts on long journeys, it has been suggested, will suffer the symptoms that afflict men who are cut off from their environment by being shut up in darkened, soundproofed rooms.

I would reverse this argument; our entire culture will suffer from sensory deprivation if it does *not* go out into space. There is striking evidence for this in what has already happened to the astronomers and physicists. As soon as they were able to rise above the atmosphere, a new and often surprising universe was opened up to them, far richer and more complex than had ever been suspected from ground observations. Even the most enthusiastic proponents of space research never imagined just how valuable satellites would actually turn out to be, and there is a profound symbolism in this.

But the facts and statistics of science, priceless though they are, tell only part of the story. Across the seas of space lie the new raw materials of the imagination, without which all forms of art must eventually sicken and die. Strangeness, wonder, mystery, adventure, magic—these things, which not long ago seemed lost forever, will soon return to the world. And with them, perhaps, will come again an age of sagas and epics such as Homer never knew.

Though we may welcome this, we may not enjoy it, for it is never easy to live in an age of transition—indeed, of revolution. As the old Chinese curse has it: "May you live in interesting times," and the twentieth century is probably the most "interesting" period that mankind has ever known. The psychological stress and strains produced by astronautics—upon the travelers and those who stay at home—will often be unpleasant, even though the ultimate outcome will be beneficial to the race as a whole.

The American public has already experienced some emotional highs and lows that give a slight foretaste of what is to come. To date, the extremes are well represented by the explosion of the first Vanguard, and the success of the first manned sub-orbital shot, when the whole nation stopped its work and play to watch Cape Kennedy. But these are only pale shadows of such future triumphs and disasters as the landing on the Moon—or the impact of a Nova-class vehicle on Miami Beach.

We must also prepare ourselves for the probability —in fact, the virtual certainty—that the most painful and uncomfortable shocks will involve our philosophical and religious beliefs. Many optimistic apologists have tried to deny this, but the clear verdict of history is against them.

We now take it for granted that our planet is a tiny world in a remote corner of an infinite universe, and have forgotten how this discovery shattered the calm certainties of medieval faith. Even the echoes of the second great scientific revolution are now swiftly fading; today, except in a few backward regions, the theory of evolution arouses as little controversy as the statement that the Earth moves round the Sun. Yet it is only a hundred years since the best minds of the Victorian age tore themselves asunder because they could not face the facts of biology.

Space will, sooner or later, present us with facts that are much more stubborn, and even more disconcerting. There can be little reasonable doubt that, ultimately, we will come into contact with races more intelligent than our own. That contact may be one-way, through the discovery of ruins or artifacts; it may be two-way, over radio or laser circuits; it may even be face to face. But it will occur, and it may be the most devastating event in the history of mankind. The rash assertion that "God made man in His own image" is ticking like a time bomb at the foundations of many faiths, and as the hierarchy of the universe is disclosed to us, we may have to recognize this chilling truth: if there are any gods whose chief concern is man, they cannot be very important gods.

The best examination I have seen of the probable effects of space travel upon our philosophical-religious beliefs was made in a broadcast by Derek Lawden, well known for his work on interplanetary orbits. Because few people outside New Zealand will have heard his stimulating talk, it is worth giving Professor Lawden's conclusions at some length:

I think man will see himself as one agent by which the whole universe of matter is slowly becoming conscious of itself. He will cease to feel an alien creature in an indifferent world, but will sense within himself the pulse of the cosmos. He'll become familiar with the marvellous and varied form which can be assumed by matter . . . and he's certain to develop a feeling of reverence for the awe-inspiring whole of which he's a very small part. I suggest to you that his reaction to these impressive experiences will find its expression in a pantheism which will at last provide a philosophy of life and an attitude to existence which is in harmony with science. . . . It may be objected that the physical universe could never become the object of worship. I ask anyone who denies this possibility to turn his eyes skyward on a clear night. . . . Others may object that such a religion would possess little moral content. I would reply that this is by no means self-evident, but that, in any case, the conjunction of religion and ethics . . . is certainly not invariable; in fact, there's an excellent case for keeping the two separate. . . . Morality in the modern Western world has been greatly weakened because of its strong ties with Christianity, for as one decays, so does the other. . . .

These are hard sayings, which many will find unpalatable; the truth may be yet harder. Perhaps if we knew all that lay ahead of us on the road to space—a hundred or a thousand or a million years in the future —no man alive would have the courage to make the first step. But that first step—and the second—has already been taken; to turn back now would be treason to the human spirit, even though our feet must some day carry us into realms no longer human.

The eyes of all the ages are upon us now, as we create the myths of the future at Cape Kennedy and Baikonur. No other generation has been given such powers, and such responsibilities. The impartial agents of our destiny stand on their launching pads, awaiting our commands. They can take us to that greater Renaissance whose signs and portents we can already see, or they can make us one with the dinosaurs.

The choice is ours, it must be made soon, and it is irrevocable. If our wisdom fails to match our science, we will have no second chance. For there will be none to carry our dreams across another dark age, when the dust of all our cities incarnadines the sunsets of the world.

The Uses of the Moon

The two greatest nations in the world are now preparing to land men on the Moon within the next decade. This will be one of the central facts of political life in the years to come; indeed, it may soon dominate human affairs. It is essential, therefore, that we understand the importance of the Moon in our future; if we do not, we will be going there for the wrong reasons, and will not know what to do when we arrive.

Many people imagine that the whole project of lunar exploration is merely a race with the Russians—a contest in conspicuous consumption of brains and material, designed to impress the remainder of mankind. No one can deny the strong element of competition and national prestige involved, but in the long run, this will be the least important aspect of the matter. If the race to the Moon were nothing more than a race, it would make good sense to let the Russians bankrupt themselves in the strain of winning it, in the calm confidence that their efforts would collapse in recriminations and purges some time during the 1970's.

There are some shortsighted people (including a few elderly, but unfortunately still influential, scientists) who would adopt just such a policy. Why spend tens of billions of dollars, they ask, to land a few men on a barren, airless lump of rock, nothing more than a cosmic slagheap, baked by the Sun during the daytime and frozen to subarctic temperatures in the long night? The polar regions of this Earth are far more hospitable; indeed, the deep oceans could probably be exploited and even colonized for a fraction of the sum needed to conquer the Moon.

All this is true; it is also totally irrelevant. The Moon *is* a barren, airless wasteland, blasted by intolerable radiations. Yet a century from now it may be an asset more valuable than the wheatfields of Kansas or the oil wells of Oklahoma. And an asset in terms of actual hard cash—not the vast imponderables of adventure, romance, artistic inspiration and scientific knowledge. Though, ultimately, these are the only things of real value, they can never be measured. The conquest of the Moon, however, can be justified to the cost accountants, not only to the scientists and the poets.

Let me first demolish, with considerable pleasure, one common argument for going to the Moon—the military one. Some ballistic generals have maintained that the Moon is "high ground" that could be used for reconnaissance and bombardment of the Earth. Though I hesitate to say that this is complete nonsense, it is as near to it as makes very little practical difference.

You cannot hope to see as much from 250,000 miles away as from a TV satellite just above the atmosphere, and the use of the Moon as a launching site makes even less sense. For the effort required to set up one lunar military base with all its supporting facilities, at least a hundred times as many bases could be established on Earth. Also it would be far easier to intercept a missile coming from the Moon, and taking many hours for the trip in full view of telescopes and radar, than one sneaking round the curve of the Earth in twenty minutes. Only if, which heaven forbid, we extend our present tribal conflicts to the other planets will the Moon become of military importance.

Before we discuss the civilized uses of our one natural satellite, let us summarize the main facts about it. They may be set down quite briefly:

The Moon is a world a quarter the diameter of Earth, its radius being just over a thousand miles. Thus its area is one-sixteenth of our planet's—more than that of Africa, and almost as much as that of both the Americas combined. Such an amount of territory is not to be despised; it will take many years (and many lives) to explore it in detail.

The amount of material in the Moon is also impressive; if you would like it in tons, the figure comes to 750,000,000,000,000,000,000,000,000, which is millions of millions of times more than all the coal, iron, minerals and ores that man has shifted in the whole of history. It is not enough mass, however, to give the Moon much of a gravitational pull; as everyone now knows, a visitor to the Moon has only a fraction (actually one-sixth) of his terrestrial weight.

This low gravity has several consequences, almost all of them good. The most important is that the Moon has been unable to retain an atmosphere; if it ever had one, it long ago escaped from the Moon's feeble clutch and leaked off into space. For all practical purposes, therefore, the lunar surface is in a perfect vacuum. (*This* is an advantage? Yes: we'll see why in a moment.)

Because there is no atmosphere to weaken the Sun's rays, or to act as a reservoir of heat during the nighttime, the Moon is a world of very great temperature extremes. On our Earth, in any one spot, the thermometer seldom ranges over as much as a hundred degrees even during the course of a year. Though the temperature can exceed 100°F in the tropics, and drop to 125° *below* zero in the Antarctic, these figures are quite exceptional. But every point on the Moon undergoes twice this range during the lunar day; indeed, an explorer could encounter such changes within seconds, merely by stepping from sunlight into shadow or vice versa.

This obviously presents problems, but the very absence of atmosphere which causes such extremes also makes it easy to deal with them—for a vacuum is one of the best possible heat insulators, a fact familiar to anyone who has ever taken hot drinks on a picnic.

No air means no weather. It is hard for us, accustomed to wind and rain, cloud and fog, hail and snow to imagine the complete absence of all these things. None of the meteorological variations which make life interesting, unpredictable and occasionally impossible on the surface of this planet takes place on the Moon: the only change which ever occurs is the regular, utterly unvarying cycle of day and night. Such a situation may be monotonous but it simplifies, to an unbelievable extent, the problems facing architects, engineers, explorers and indeed everyone who will ever conduct operations of any kind on the surface of the Moon.

The Moon turns rather slowly on its axis, so that its day (and its night) are almost thirty times longer than ours. As a result, the sharp-edged frontier between night and day, which moves at a thousand miles an hour on the Earth's equator, has a maximum speed of less than ten miles an hour on the Moon. In high lunar latitudes, a walking man could keep in perpetual daylight with little exertion. And because the Moon turns on its axis in the same time as it revolves around the Earth, it always keeps the same hemisphere turned toward us. Until the advent of Lunik III, this was extremely frustrating to astronomers; in another generation, as we shall see, they will be very thankful for it.

So much for the main facts; now for a few assumptions which most people would accept as reasonable in 1961, though they would have laughed at them before 1957.

The first is that suitably protected men can work and carry out engineering operations on the face of the Moon, either directly or by remote control through robots.

The second is that the Moon consists of the same elements as the Earth, though doubtless in different proportions and combinations. Most of our familiar minerals will be missing: there will be no coal or limestone, since these are the products of life. But there will be carbon, hydrogen, oxygen, and calcium in other forms, and we can evolve a technology to extract them from whatever sources are available. It is even possible that there may be large quantities of free (though frozen) water not too far below the Moon's surface; if this is the case, one of the chief problems of the lunar colonists will be solved.

In any event, without going into details of mining, ore processing and chemical engineering, it will be possible to obtain all the materials needed for maintaining life. The first pioneers will be content with mere survival, but at a later stage they will build up a self-supporting industry based almost entirely on lunar resources. Only instruments, specialized equipment and men will come from Earth; the Moon will supply all the rest—ultimately, of course, even the men.

There have been many studies and books on the subject of lunar colonization (I have written one myself) and all those who have gone into the subject are agreed on the general picture. The details vary, as they must until we have much more exact knowledge of conditions on the Moon, but that is of no importance. It may take as little as fifty years (the interval between the Wright biplane and the B-52!) to establish a viable lunar colony; it may take a hundred. But if we wish, it can be done; on the Moon, to borrow the words of William Faulkner's Nobel Prize speech, "Man will not merely survive—he will prevail."

Now for the reasons why it is worth the expense, risk and difficulty of prevailing on the inhospitable Moon. They are implicit in the question: what can the Moon offer that we cannot find on Earth?

One immediate but paradoxical answer is Nothing—millions of cubic miles of it. Many of the key industries in the modern world are based on vacuum techniques;

electric lighting and its offspring radio and electronics could never have begun without the vacuum tube, and the invention of the transistor has done little to diminish its importance. (The initial steps of transistor manufacture have themselves to be carried out in vacuum.) A great many metallurgical and chemical processes, and key stages in the production of such drugs as penicillin, are possible only in a partial or virtually complete vacuum; but it is expensive to make a very good vacuum, and impossible to make a very large one.

On the Moon, there will be a "hard" vacuum of unlimited extent outside the door of every airlock. I do not suggest that it will be worthwhile switching much terrestrial industry to the Moon, even if the freight charges allowed it. But the whole history of science makes it certain that new processes and discoveries of fundamental importance will evolve as soon as men start to carry out operations in the lunar vacuum. Low-pressure physics and technology will proceed from rags to riches overnight; industries which today are unimagined will spring up on the Moon and ship their products back to Earth. For in that direction, the freight charges will be relatively low.

And this leads us to a major role that the Moon will play in the development of the solar system: it is no exaggeration to say that this little world, so small and close at hand (the very first rocket to reach it took only thirty-five hours on the journey) will be the stepping stone to all the planets. The reason for this is its low gravity; it requires twenty times as much energy to escape from the Earth as from the Moon. As a supply base for all interplanetary operations, therefore, the Moon has an enormous advantage over the Earth— assuming, of course, that we can find the materials we need there. This is one of the reasons why the development of lunar technology and industry is so important.

From the gravitational point of view, the Moon is indeed high ground, while we on the Earth are like dwellers at the bottom of an immensely deep pit out of which we have to climb every time we wish to conduct any cosmic explorations. No wonder that we must burn a hundred tons of rocket fuel for every ton of payload we launch into space—and on a one-way trip at that. For return journeys, thousands of tons would be needed.

This is why all Earth-based plans for space travel are so hopelessly uneconomic, involving gigantic boosters with tiny payloads. It is as if, in order to carry a dozen passengers across the Atlantic, we had to construct a ship weighing as much as the *Queen Elizabeth* but costing very much more. (The development costs for a large space vehicle are several billion dollars.) And, to make the whole thing completely fantastic, the vehicle can be used only once, *for it will be destroyed in flight*. Of the tens of thousands of tons that leave the Earth, only a small capsule will return. The rest will consist of boosters dropped into the ocean or discarded in space.

When nuclear power is harnessed for rocket propulsion, the position will be improved from the preposterous to the merely absurd. For even nuclear rockets must carry hundreds or thousands of tons of reaction mass, to provide a thrust when it is ejected. Every rocket, nuclear or chemical, has to have something to push against; that something is not the surrounding air, as many people once believed, but the rocket's own fuel.

However, the nuclear rocket will use the very simplest of fuels—plain hydrogen. There must be plenty of this on the Moon, combined in water (which is 11 per cent hydrogen) or in some other form. The first order of business in lunar exploration will be to locate sources from which hydrogen may be obtained; when this has been done, and it is possible for ships to refuel on the Moon, the cost, difficulty and complexity of all space operations will be reduced at least tenfold.

Since spacecraft need not carry fuel for the return trip (imagine where transatlantic flying would be today, if it operated on this basis!) it will no longer be necessary to build and jettison ten-thousand-ton vehicles to deliver ten ton payloads. Instead of monstrous, multistaged boosters, we can use relatively small rockets that can be refueled and flown over and over again. Space flight would emerge from its present status as a fantastically expensive stunt, and would start to make economic—perhaps even commercial—sense.

This, however, would be only a beginning. The big breakthrough toward really efficient space operations may depend upon the fortunate fact that the Moon has no atmosphere. The peculiar (by our standards—they are normal by those of the universe) conditions prevailing there permit a launching technique much more economical than rocket propulsion. This is the old idea of the "space gun," made famous by Jules Verne almost a hundred years ago.

It would probably not be a gun in the literal sense, powered by chemical explosives, but a horizontal launching track like those used on aircraft carriers, along which space vehicles could be accelerated electrically until they reached sufficient speed to escape from the Moon. It is easy to see why such a device is completely impractical on Earth, but might be of enormous value on the Moon.

To escape from the Earth, a body must reach the now familiar speed of 25,000 miles an hour. At the fierce acceleration of ten gravities, which astronauts have already withstood for very short periods of time, it would take two minutes to attain this speed—and the launching track would have to be *four hundred miles* long. If the acceleration were halved to make it more endurable, the length of the track would have to be doubled. And, of course, any object traveling at such a speed in the lower atmosphere would be instantly burned up by friction. We can forget all about space guns on Earth.

The situation is completely different on the Moon. Because of the almost perfect vacuum, the lunar escape speed of a mere 5,200 m.p.h can be achieved at ground level without any danger from air resistance. And at an acceleration of ten gravities, the launching track need be only 19 miles long—not 400, as on the Earth. It would be a massive piece of engineering, but a perfectly practical one, and it would wholly transform the economics of space flight.

Vehicles could leave the Moon *without burning any fuel at all;* all the work of take-off would be done by fixed power plants on the ground, which could be as large and massive as required. The only fuel that a space vehicle returning to Earth need carry would be a very small amount for maneuvering and navigating. As a result, the size of vehicle needed for a mission from Moon to Earth would be reduced tenfold; a hundred-ton spaceship could do what had previously required a thousand-tonner.

This would be a spectacular enough improvement; the next stage, however, would be the really decisive one. This is the use of a Moon-based launcher or catapult to place supplies of fuel where they are needed, in orbit round the Earth or indeed any other planet in the solar system.

It is generally agreed that long-range space flight—particularly voyages beyond the Moon—will become possible only when we can refuel our vehicles in orbit. Plans have been drawn up in great detail for operations involving fleets of tanker rockets which, perhaps over a period of years, could establish what are virtually filling stations in space. Such schemes will, of course, be fantastically expensive, for it requires about fifty tons of rocket fuel to put a single ton of payload into orbit round the Earth, only a couple of hundred miles up.

Yet a Moon-based launcher could do the same job —from a distance of 250,000 miles!—for a twentieth of the energy and without consuming any rocket fuel whatsoever. It would launch tanks of propellants "down" toward Earth, and suitable guidance systems would steer them into stable orbits where they would swing around endlessly until required. This would have as great an effect on the logistics of space flight as the dropping of supplies by air has already had upon polar exploration; indeed, the parallel is a very close one.

Though enormous amounts of power would be required to operate such lunar catapults, this will be no problem in the twenty-first century. A single hydrogen bomb, weighing only a few tons, liberates enough energy to lift a hundred million tons completely away from the Moon. That energy will be available for useful purposes when our grandchildren need it; if it is not, we will have no grandchildren.

There is one other application of the lunar catapult that may be very important, though it may seem even more farfetched at the present time. It could launch the products of the Moon's technology all the way down to the surface of the Earth. A rugged, freight-carrying capsule, like a more refined version of today's nose cones and re-entry vehicles, could be projected from the Moon to make an automatic landing on the Earth at any assigned spot. Once again, no rocket fuel would be needed for the trip, except a few pounds for maneuvering. All the energy of launching would be provided by the fixed power plant on the Moon; all the slowing down would be done by the Earth's atmosphere. When such a system is perfected, it may be no more expensive to ship freight from Moon to Earth than it is now to fly it from one continent to another by jet. Moreover, the launching catapult could be quite short, since it would not have to deal with fragile human passengers. If it operated at fifty gravities acceleration, a four-mile-long track would be sufficient.

I have discussed this idea at some length for two reasons. The first is that it demonstrates how, by taking advantage of the Moon's low gravity, its airlessness, and the raw materials that must certainly be there, we can conduct space exploration far more economically than by basing our operations on Earth. In fact, until some revolutionary new method of propulsion is invented, it is hard to see any other way in which space travel will be practical on the large scale.

The second reason is the slightly more personal one

that, to the best of my knowledge, I was the first to develop this idea in a 1950 issue of the *Journal* of the British Interplanetary Society. Five years earlier I had proposed the use of satellites for radio and TV communications; I did not expect to see either scheme materialize in my lifetime, but one has already happened and now I wonder if I may see both.

The subject of communications leads us to another extremely important use of the Moon. As civilization spreads throughout the solar system, the Moon will provide the main link between Earth and her scattered children. For though it is just as far to the other planets from the Moon as from the Earth, sheer distance is not the only factor involved. The Moon's surface is already in space, while the surface of the Earth—luckily for us—is shielded from space by a whole series of barriers through which we have to drive our signals.

The best known of these barriers is, of course, the ionosphere, which reflects all but the shorter radio waves back to Earth. The shortest waves of all, however, go through it with little difficulty, so the ionosphere is no hindrance to space communications.

What *is* a serious barrier—and this has been realized only during the past year—is the atmosphere itself. Thanks to the development of an extraordinary optical device called the laser, which produces an intense beam of almost perfectly parallel light, it now appears that the best agent for long distance communications is not radio, but *light*. A light beam can carry millions of times as many messages as a radio wave, and can be focused with infinitely greater accuracy. Indeed, a laser-produced light beam could produce a spot on the Moon only a few hundred feet across, where the beam from a searchlight would be thousands of miles in diameter. Thus colossal ranges could be obtained with very little power; calculations show that with lasers we can think of signaling to the stars, not merely to the planets.

But we cannot use light beams to send messages through the Earth's erratic atmosphere; a passing cloud could block a signal that had traveled across a billion miles of space. On the airless Moon, however, this would be no problem, for the sky is perpetually clear to waves of all frequencies, from the longest radio waves, through visible light, past the ultraviolet and even down to the short X rays which are blocked by a few inches of air. This whole immense range of electromagnetic waves will be available for communications or any other use—perhaps such applications as the broadcasting of power, which have never been practical on Earth. There will be enough "band width" or ether space for all the radio and TV services we can ever imagine, no matter how densely populated the planets become and however many messages the men of the future wish to flash back and forth across the solar system.

We can thus imagine the Moon as a sort of central clearing house for interplanetary communications, aiming its tightly focused light beams to the other planets and to ships in space. Any messages that concerned Earth would be radioed across the trivial 250,000 mile gulf on those wavelengths that penetrate our atmosphere.

There are several other reasons why the Moon might almost have been designed as a base for interplanetary communications. Everyone is now familiar with the enormous radio telescopes which have been built to reach out into space and to maintain contact with such distant probes as our Pioneers and Explorers (and the Rangers, Mariners and Prospectors that will follow them). The most ambitious of these was the ill-fated 600-foot giant at Sugar Grove, West Virginia—abandoned when partly built, after some scores of millions of dollars had been spent on it.

The 600-foot telescope was an expensive failure because it was too heavy; the planned weight was about 20,000 tons, but design changes later brought it up to 36,000 tons. But on the Moon, both the cost and weight of such a structure would be enormously reduced—perhaps by more than 90 per cent. For thanks to the low gravity, a very much lighter construction could be used than is necessary on Earth. And the Moon's airlessness pays another dividend, for a terrestrial telescope has to be designed with a substantial safety factor so that it can withstand the worst that the weather can do. There is no need to worry about gales on the Moon, where there is not the slightest breeze to disturb the most delicate structures.

Nor have we yet finished with the Moon's advantages from the view of those who want to send (and receive) signals across space. It turns so slowly on its axis that the problem of tracking is much simplified; *and it is a quiet place*.

Or, to be more accurate, the far side of the Moon is a quiet place—probably the quietest that now exists within millions of miles of the Earth. I am speaking, of course, in the radio sense; for the last sixty years, our planet has been pouring an ever increasing racket into space. This has already seriously inconvenienced the radio astronomers, whose observations can be ruined by an electric shaver a hundred miles away.

But the land first glimpsed by Lunik III is beyond the reach of this electronic tumult; it is shielded from the din of Earth by two thousand miles of solid rock —a far better protection than a million miles of empty space. Here, where the earth-light never shines, will be the communications centers of the future, linking together with radio and light beams all the inhabited planets. And one day, perhaps, reaching out beyond the solar system to make contact with those other intelligences for whom the first search has already begun. That search can hardly hope for success until we have escaped from the braying of all the radio and TV stations of our own planet.[1]

What has already been said should be enough to convince any imaginative person—anyone who does not believe that the future will be a carbon copy of the past—that the Moon will be a priceless possession and

---

[1] I am happy to report that the 1963 Geneva conference of the International Telecommunications Union made specific reference to the need for protecting the lunar far side from interference.

its exploration far more than the expensive scientific stunt that some foolish people have called it. At the same time it should be emphasized that the most important and valuable uses of the Moon will be ones that nobody has thought of today. This has always been true in the past of exploration and scientific discovery, and it will be equally true in the future. I will merely hint at a few possibilities here.

In a recent discussion of space exploration plans, Professor Harold Urey made the point that the Moon is one of the most interesting places in the solar system—perhaps more so than Mars or Venus, even though there may be life on these planets. For the face of the Moon may have carried down through the ages, virtually untouched by time, a record of the conditions that existed billions of years ago, when the universe itself was young. On Earth, all such records have long been erased by the winds and the rains and other geological forces. When we reach the Moon, it will be as if an entire library of lost volumes, a million times older than that destroyed at Alexandria, was suddenly thrown open to us.

Quite beyond price will be the skills we will acquire during the exploration—and ultimately, colonization—of this new land in the sky. I suspect, though only time will tell whether this is true, that we will learn more about unorthodox methods of food production on the Moon within a few years than we could in decades on the Earth. Can we, in an almost literal sense of the phrase, turn rocks into food? We must master this art (as the plants did, aeons ago) if we hope to conquer space. Perhaps most exciting of all are the possibilities opened up by low-gravity medicine and the enormous question: "Will men live longer on a world where they do not wear out their hearts fighting against gravity?" Upon the answer to this will depend the future of many worlds, and of nations yet unnamed.

Much of politics, as of life, consists of the administration of the unforeseen. We can foresee only a minute fraction of the Moon's potentialities, and the Moon itself is only a tiny part of the universe. The fact that the Soviet Union is making an all out effort to get there has far deeper implications than have been generally faced.

The Russians, whatever else they may be, are realists. And as Sir Charles Snow has pointed out in his highly influential book, *Science and Government,* between 35 and 45 per cent of their top men have some technical and scientific training. (It is doubtful if the proportion is a quarter of this in the West.) As a result, they have often made correct choices—for example, the decision to develop the lithium bomb and giant rocket boosters —when the United States wasted its energies in such technological dead ends as tritium bombs and air-breathing missiles.

They may have done so again in the most important field of all. I wonder if any of the "Leave it to the Russians" school of anti-space flight critics seriously imagines that Soviet science is outward bound merely to impress the uncommitted nations. That could be achieved in a dozen less expensive ways.

No, the Russians know exactly what they are doing. Perhaps they are already laughing at the shortsighted prophets who have said: "Anyone who owns the Moon can dominate the Earth."

They may no longer be concerned with such trivialities. They realize that if any nation has mastery of the Moon, it will dominate not merely the Earth, but the whole accessible universe.

If, in November 1967, there are only Russians on the Moon to drink a toast to the fiftieth anniversary of the Revolution, they will have won the solar system, and theirs will be the voice of the future.

As it will deserve to be.

Ours is the first generation, in the whole history of mankind, that must learn to think in millions of miles. Until the dawn of the space age, there were no journeys of more than a thousand miles; indeed, to most of our ancestors, a hundred miles was the limit of a lifetime's traveling. Only sailors, merchants, soldiers, and explorers were concerned with anything beyond this range.

This was still true until well into the nineteenth century, when improvements in communication and the rise of literacy made even the most stay-at-home aware of the whole Earth, as an eight-thousand-mile-diameter globe floating in space. Today, every educated person has a clear mental picture of that globe and of the distances that are involved in moving around it. Any intelligent schoolboy is now familiar with a world almost fifty times greater than that known to a cultured, much-traveled genius like Julius Caesar.

The story of civilization is, indeed, largely one of expanding mental horizons. First there was the village; then the town; then the province; then the country; then the empire; then the world. Now there is the solar system—and we must learn to grasp it with our minds, as our ancestors grasped this little Earth with theirs.

It is sometimes said that astronomical distances are so enormous that they are meaningless—beyond all imagination or understanding. This is a misleading half-truth; though no one may be able to visualize the distances that separate the planets, it is not difficult

to become familiar with them. After all, we have to
handle millions and even billions in everyday life—
though no one can actually hold a mental picture of a
number much larger than five or six. (Nine is about
my limit—I can just visualize it by imagining three
rows of three objects. If you can do better than that,
you have my skeptical admiration.)

For that matter, can anyone *really* picture a distance
of a thousand miles—of a hundred miles—or even of
ten miles? It is doubtful, yet this does not hinder us
from working with such distances. It is very much
easier to grasp the mileages of the solar system than
the expenditures of the U.S. budget; yet this last feat
is apparently performed by several hundred Congressmen who would admit a total ignorance of mathematics
and astronomy. Anyone who pretends that he cannot
cope with the universe is, therefore, just being lazy.

Very early in history men invented a trick for making large quantities manageable; they simply invented
suitable names or labels for them. At first those names
dealt with fairly modest amounts; thus we have such
words as dozen, score, gross, which are adequate for
most of the quantities met in everyday life. In the West
we seem to have stopped with gross (or a dozen dozen),
though at one time the word myriad was employed to
mean 10,000. In the East, on the other hand, a much
larger unit, the lakh (100,000) is in common use, and
is so convenient that I sometimes miss it when I am
outside the Orient.

The introduction of the metric system, at the end of
the eighteenth century, gave the world a simple and
consistent means of naming quantities of any magnitude. The prefix "kilo" for a thousand soon became
universally familiar through the kilometer and the kilo-
gram; even backward countries such as England and
the United States are forced to use the kilowatt, though
they still cling tenaciously to such outmoded units as
feet and yards, pounds and ounces.

It took somewhat longer for the prefix "mega" to
be commonly accepted outside scientific circles, and it

finally made its advent via the H-bomb in the form of the megaton. It is perfectly legitimate, though somewhat unusual, to talk about the megamile, and this is a very convenient unit for dealing with the solar system. We can write it as Mile, with a capital "M"; from now on, therefore, Mile will mean megamile or million miles, and mile with a small "m" will mean the ordinary, everyday mile.

The nearest object in space—the Moon—is just a quarter of a Mile away, and orbiting astronauts have already traveled much further than this. (Forty times around the Earth is approximately one Mile, and that number of close orbits takes less than three days.) At its closest, the planet Venus passes within twenty-five Miles of us, Mars within thirty-six, and the Sun itself is ninety-three Miles away. Manmade objects like the Mariner Venus and Mars probes have now traveled more than two hundred Miles from their parent planet.

As we go further away from the Sun, the scale of the solar system becomes larger, and the gulfs between the planets widen. Thus the gap Mars-Jupiter is 340 Miles, Jupiter-Saturn 400 Miles, Saturn-Uranus 900 Miles, Uranus-Neptune 1,000 Miles. These are the shortest distances between the orbits; the planets themselves are normally much further apart, as they pursue their separate paths around the Sun. Pluto is omitted from the list because its orbit is an abnormal one; though it is the most distant planet, its orbit is so eccentric that it sometimes comes inside that of Neptune. During the closing years of this century Neptune, not Pluto, will be the farthest planet from the Sun.

For a good many years to come, all our manned explorations will be concerned with the region inside Mars—that is, a disk of space about 250 Miles across. Automatic probes, however, will be ranging much further afield—indeed, over the whole solar system. If we consider that the limits of the solar system are set by the orbit of Pluto, this is an area about seven thousand Miles across.

All these numbers, you will note, are about the same

size as those we use when discussing distances on the Earth; in other words, solar system distances are just about a million times as big as terrestrial ones. But from the point of view of the traveler, it is time, not space, that is important. If we want to move around the solar system as swiftly as we now do around the world, we must attain speeds of several hundred Miles an hour.

At the moment, our rockets cannot achieve a thousandth—or even a *ten* thousandth—of this enormous speed. Even if space vehicles could be developed to reach it, such velocities would be somewhat ill-advised in such a—relatively!—crowded neighborhood as the solar system. Space between the planets is by no means empty; meteorites, cosmic dust, comets and minor planets exist in numbers which would probably be a severe hazard to super-high-speed interplanetary flight. A collision with even a very small body at a velocity of a few Miles an hour would release as much energy as an atomic bomb.

However, we do not need super speeds in order to reach the nearer planets in a reasonable time. The first explorers of the Earth managed without jets; they had to travel on foot, and it took them months or years to make continent-wide journeys. Our first space explorations will be on much the same time scale; in the early days of interplanetary flight, the one-way voyage to Venus will take about five months, that to Mars about eight. As long as we are restricted to chemical fuels, we will never be able to do much better than this.

But to imagine that today's—or even tomorrow's— rockets represent the last word in space technology is as ridiculous as thinking that a war chariot is the fastest land vehicle that will ever be built. The Babylonians and the Assyrians may well have believed this, and it was true for a good many thousand years. However, progress is slightly more rapid nowadays; as far ahead as we can imagine, every spaceship will be already obsolete as soon as it goes into service.

During the next century, as we learn to harness

nuclear power for space propulsion, the solar system will contract as the Earth itself has shrunk since men began to explore it. Only a century ago Jules Verne astonished everyone by suggesting that a man might go round the world in eighty days; fifty-six years after Verne's death, Gagarin had done it in little more than eighty minutes. The time may come—perhaps before 2100 A.D.—when men will think no more of traveling to Mars or the satellites of Jupiter than they do today of jetting across the Pacific. And the flight to the Moon will be too unimportant to mention.

Yet the far-flung solar system—the Sun with its family of planets, satellites, minor planets and comets—is a tiny island in an immense and empty ocean. Just as we have had to stretch our minds a million times to grasp the distances of the planets, so we will have to stretch them *another* millionfold to grasp the distances of the stars. Even those who have made the first step may well balk at the second, but unless it is taken, we cannot come to grips with the scale and proportions of the universe.

We invented the megamile or Mile to handle the solar system; let us try the same trick again. The mega-megamile (a million million or trillion miles) could be written MMile, and in this unit the distance to the nearest star (Proxima Centauri) is 25 MMiles. The brightest star, Sirius, is about 50 MMiles away. To give some idea of how the stars are scattered through space, there are about 50 within 100 MMiles of the Sun. Incidentally, our Sun is quite a respectable star; though far brighter and larger stars exist, the great majority are much smaller and dimmer than the one that warms our planet.

Though it is now obvious, even to the most conservative and unimaginative, that the planetary distances will be of practical importance in the quite near future, there are some who would argue that the millionfold greater distances to the stars are of no concern to anyone except the astronomers. This is simply not true, for several excellent reasons.

First of all, every intelligent, cultured person should have some idea of the correct size of the universe in which he lives. We would not think much of a New Yorker who imagined that Australia was the next stop past Staten Island—yet even this ridiculous example scarcely exaggerates the ignorance of many people regarding the true size of space. In particular, even educated men are often quite unable to grasp the difference between the *planetary* distances which we are challenging now, and the *stellar* ones which separate us from those other suns, the stars. One cannot blame them; a change in scale of a million to one is very hard to conceive and, as we have seen, it has to be made not once but twice if we wish to raise our sights from Earth to the planets, and then out to the stars.

Yet we must make the effort; one day even the stars, despite their immense distances, will be the direct concern of governments and politicians—and the time may be sooner than we think. Very few of the men who now sit in Congress could ever have imagined, when they went to school, that it would one day be essential for them to know something about Moon and planets, about orbits and satellites. The idea would have seemed utterly fantastic—until the fateful hour when Sputnik I soared into space.

Yet by a strange coincidence, the stars and the planets—despite the enormous disparity in their distances—have both come within our range at the same time. Though we cannot *physically* make contact with the stars, as we shall soon do with the planets, we may soon be able to make mental contact with them.

Today, no one expects that we will find intelligent life on the other planets of this solar system—though almost everyone expects that we will discover some form of life on Mars. We may, of course, be pleasantly (or even unpleasantly) surprised when we get to Mars; but all the evidence points to the fact that our little red neighbor is a very much older planet than Earth, and any intelligent Martians must have become extinct millions of years ago. The best we can hope to do is to find their fossils, and the buried ruins of their cities, but even this is pure guesswork. It is much more likely that Earth is the only one of the Sun's children that has ever given rise to intelligence.

If we find that we are alone in the solar system (and this is something that we shall know before the end of this century), it will have a profound effect upon our philosophical outlook. This will be even truer if we discover life—but only *unintelligent* life—on the other planets. Such a discovery would be both exciting and frustrating; it would increase our feeling of isolation in this enormous universe, and at the same time make it all the more probable that elsewhere, on planets circling other stars, there must be forms of life that possess intelligence.

At first sight, it may appear impossible that we could ever learn anything about such beings, megamegamiles away. Yet though Proxima Centauri is a million times further away than Mars, it is already nearer to us, in terms of our space technology, than Mars was to the first astronomers who observed it through their crude telescopes, three hundred years ago.

It is a surprising fact that the space probes we shall build in the next generation will be able to reach the nearer stars. To escape from the Earth, a rocket must attain the well-known "velocity of escape," which is about seven miles a second or 25,000 miles an hour. There is also a "velocity of escape" not only from the Earth but from the solar system—and it works out, in our neighborhood, at an astonishingly low ten miles a second. Thus a 50 per cent increase in speed will allow our present-day Mariners and Rangers to get to the stars! And if they were fitted with suitable guidance mechanisms and a modest ability to change course by low-powered control jets, they could make a survey of the star they had approached, whip round it in a comet-like orbit, and head back to Earth again, without burning up more than a few pounds of rocket fuel. What a cargo of knowledge they could bring back to the eagerly waiting scientists!

Unfortunately, there is a slight catch. True, tomorrow's rockets *could* make the trip—but it would last about a million years. Scientists are patient men, but we can hardly expect them to wait that long. If the earliest ape-men had spent their time launching interstellar probes instead of chipping flints, we might expect the first ones back by now.

Of course, as space-vehicle performances improve, these transit times will be reduced drastically. With a tenfold increase in speed, it would take "only" a hundred thousand years to make the trip. But even if we assume that there is rapid progress in the development of space-propulsion systems (as there certainly will be), it does not follow that we will soon launch probes to the stars. In fact, the more rapid the progress, the *less* likely such a launching would become!

This sounds ridiculous, but a little thought will soon show why it must be so, in this universe of ours. Let us survey, in imagination, the fortunes of Project Centauri, an effort started by the National Aeronautics and Space Administration around the year 2000 to send a space probe to the nearest star.

By that time, we will assume, it is possible to design a probe that can make the round trip to Proxima Centauri in a hundred thousand years. It has taken half a century of space research to make such a feat possible —and, obviously, it is not worth doing. No one can be sure that the human race, still less NASA, will be around in 102000 A.D. to look at the results.

Besides, there is a much better reason for postponing action. Give us another hundred years, say the scientists, and we'll be able to make a probe that is ten times faster still. It will do the trip in a mere ten thousand years—and will be back in the year 12100 instead of 102000. This is obviously a much more practical proposition, so the building of the hardware is put off for a century.

However, when the year 2100 arrives, there will have been some more breakthroughs. The scientists *can* build the ten-thousand-year voyager, if anyone wants it, but they are also sure that by the end of the century they can improve performance another tenfold. A probe, backed by the technology of the year 2200, will be able to make the trip in only a thousand years; it will be home again by the year 3200.

In other words, the more rapid the rate of technical development, the better it is to wait before committing oneself to actual operations. In the case of a probe to the stars, we are quite sure that *no* improvement in technology will ever enable us to build a rocket which can make the Earth-Proxima Centauri-Earth trip in less than ten years. To do this, it would have to reach almost the velocity of light—the limiting speed which cannot be exceeded by any object in our universe. This sets an ultimate ceiling for performance; when that ceiling comes in sight, it is time to stop planning and to start building.

It is anyone's guess how soon we will be able to build space probes that can make round flights to the stars in a few decades. It may even become possible during the next century; it will certainly become possible *some* day. A peaceful and prosperous global society, no longer forced to spend billions every year on preparations for war, might well start launching deep-space probes to the nearer stars as soon as there was a reasonable chance of getting them back within a few hundred years.

However, long before interstellar *probes* become feasible, we may have interstellar *communication*. It is far easier to send a radio signal to the stars than a machine—and also much quicker, since the signal will always travel at the speed of light. The astonishing developments in radio during the past twenty years, culminating in the maser and the giant radio telescopes that now sweep the skies, have given us the technology needed to send messages to the nearer stars—and to receive any that may be sent to us.

The great question, of course, is whether there is anyone out there to send or receive messages. This is now the subject of furious discussion among the astronomers and space scientists, but it can never be decided by argument alone. The only way to find the answer is to start listening, and attempts to do this have already been made. No one expects success for many years; as one astronomer put it, the chance of success is almost zero—but unless we try, it will be exactly zero. And because it is hard to think of anything more important than making contact with another intelligent race, the experiment is worth attempting even against enormous odds.

Calculations show that we could, with considerable effort and expense, build radio transmitters which could send detectable signals to several of the nearer stars. Since we have known radio for only half a century, and our whole electrical industry is not very much older, we may be about the most backward communications engineers in the whole Galaxy. If there are other intelligences out in space, they may look on our efforts with the amused contempt that we reserve for smoke signals and semaphores. So at this stage in our civilization, it would be advisable for us to listen first, rather than to attempt to talk to the stars ourselves. It will be very much cheaper and easier—and, of course, until we do start picking up signals, we will have no idea in which direction to aim our own beams.

Even if we have talkative neighbors only next door, orbiting Proxima Centauri, it would take four and a quarter years for their messages to reach us, and another four and a quarter for our answers (assuming that we replied instantly) to get back to them. There are astronomical reasons for thinking that Proxima is an unlikely place for inhabited planets, but it seems quite possible that there may be intelligent beings within ten years' talking distance (twenty years both ways!) of us.

The time delay, inevitable with any form of inter-stellar communication, emphasizes once again the vast scale of the universe, as compared with our cozy little solar system. It takes only one and a half seconds to radio a message to the Moon; Mars and Venus are just a few minutes away—and even distant Pluto could be contacted in six hours. But it is four and a quarter years to the very nearest of the stars. . . .

And yet, as we have seen, men will be attempting to talk to the stars at about the time they will be trying to reach the planets. To put it crudely, we can shout—and listen—a million times further than we can travel.

The universe must be full of voices, calling from star to star in a myriad tongues. One day we shall join that cosmic conversation, though it may be ages before we have much to contribute to it. And perhaps it is just as well that we must talk for centuries before there is any hope of crossing the megamegamiles sundering us from our equals, and our masters.

First we must cultivate our solar system; and when that is done, we shall be ready to face the stars.

Beyond Centaurus

As we have yet to reach the Moon and planets, it may seem slightly premature to worry about flights to the million-times-more-distant stars. The exploration of the solar system will keep us busy for centuries; why bother about Alpha Centauri and points beyond?

For the best of reasons. Although the discoveries we shall make on our neighboring worlds will revolutionize our knowledge of the universe, and probably transform human society, it now seems most unlikely that we shall find *intelligent* life on the other planets of this Sun. The odds are fantastically against it, when we consider the immense vistas of geological—and astronomical—time.

The solar system was formed at least five billion years ago; modern man is less than a million years old, and his civilization stretches back for little more than five thousand years. We will not be far out if we say that our Earth is a *million* times older than the culture it now precariously carries. It is, therefore, ridiculous to hope that, only next door, we shall find creatures anywhere near our level of development at this fleeting moment of time. The Martians may have been contemporaries of the great reptiles; the Cythereans may lie even further in the future.

We shall know the truth by the end of this century, but today it appears overwhelmingly likely that we are alone in the solar system. With the exception of the dolphins, who may have better things to do, there is no one to talk to us within light-years of the Sun. And so, inevitably, farsighted scientists are turning their minds toward the stars.

Several recent developments have almost forced astronomers—some with reluctance, some with enthusiasm—to think seriously about communication with the planets of other suns. Only a generation ago, there was grave doubt that such planets existed; the still popular works of such prewar writers as Jeans and Eddington asserted that the solar system was probably unique, and that Earth might be the only abode of life in the universe.

Today, thanks to improved knowledge of astronomy and biology, the position is completely reversed. It is now believed that planetary systems are extremely common, and that life will arise inevitably on any world where it is given half a chance. How often life will evolve toward intelligence, and how often intelligence will take the path that leads to technology and science, there is no way of computing. One can only guess; and since our Galaxy contains a hundred thousand million stars, it may well hold—at this very moment—millions of societies superior to our own.

Until about 1960, such speculations were of only theoretical interest, since there seemed no way in which they could be either proved or disproved. Then came the invention of the maser, which permitted a degree of noiseless radio amplification never before possible. The communications engineers did a few sums, and arrived at an astonishing result. If anyone put up the money—and if there was someone listening at the other end—it would not be difficult to build a microwave system that could send messages to the nearer stars. And if we can do this, less than a century after we have invented the telephone, then interstellar signaling should be a trivial engineering feat for a really advanced culture.

Since it is much easier—and cheaper—to receive radio messages than to transmit them, it will be some time before we attempt to send signals to the stars; but there is no reason why we should not start listening. As is well known, such experiments have already begun; however, even the optimists behind "Project OZMA" do not expect results for decades, perhaps for centuries. The odds against success are very high, but the prize is so great that the experiment is worth attempting. If we are lucky, we may in our own lifetimes hear the first intelligent signals from outer space. At the same time, we may well be relieved to know that all the radio and TV programs yet launched from Earth have sunk below the level of cosmic noise before passing beyond the orbit of Pluto.

All these speculations, occurring simultaneously with the rising tempo of space research and preparations for manned interplanetary flight, have provided a basis for still more ambitious schemes. To talk to the stars—or even to exchange visual images, which is only slightly more difficult—will certainly be exciting, but it will also be very frustrating. Man, the inveterate explorer, will never be content with secondhand information; he will want to see for himself. In a century or so, he will have visited all the planets of *this* Sun; and, almost certainly, by advanced astronomical techniques, he will have detected planets of other stars. For every world that harbors beings capable of signaling across interstellar space, there must be many that cannot do so, yet would be well worth investigating. The societies that produced Lincoln, Shakespeare, and Socrates were scarcely primitive, yet they could not make their presence known even upon the nearby Moon. Space must be full of fascinating cultures and life forms which can be studied only by direct, physical contact.

Today, no competent person doubts that we can reach the planets; but can we reach the stars? There is now a splendid fight shaping up among the space scientists over this very question. Rumors of this battle, which so far has been conducted decorously in the pages of the technical journals, have as yet scarcely reached the public, and whatever its outcome no one expects that NASA will be providing funds for interstellar flight in the foreseeable future. Yet this is no esoteric controversy between specialists; it affects our entire outlook upon the universe, upon our place in it, and, conceivably, upon our origin.

For though we may be centuries from achieving interstellar travel, if it is possible, someone must have done it already. And not once, but many, many times in the history of our huge and ancient Galaxy. How often have we had visitors in the past? How often may we expect them in the future? These are not frivolous questions; their answers may shake our civilization to its very roots.

Many scientists are so appalled by the sheer size of the universe that they flatly deny the possibility of flight to the stars. Their attitude has been breezily defined by the Harvard radio astronomer Edward Purcell as follows: "All this stuff about traveling around the universe in space suits—except for *local* exploration—belongs back where it came from, on the cereal box." Similar views have been expressed by other eminent scientists who have looked into the mathematics of interstellar rocketry. To anyone who, like myself, spent most of the thirties and all of the forties trying to convince people that we *could* fly to the Moon, such negative predictions have a depressingly familiar ring. And they are just as ill-founded as the assertions—

remember them?—that man would always be confined to the planet of his birth. The remark that "the only thing we learn from history is that we learn nothing from history" is sometimes as true of scientists as of statesmen.

Almost forty years ago, the British physicist J. D. Bernal pointed out, in a brilliantly imaginative booklet, *The World, the Flesh and the Devil,* that flight to the stars would be possible by the use of self-contained "Space Arks," virtually miniature worlds which could make voyages lasting thousands of years. Generations would live and die aboard them, knowing no other existence until the voyage drew to its end. The building of such vessels would not be an impossible task for an advanced, stable society, and if there is no other way of exploring the universe, this is how it will be done. A variation on this theme is the shipload of deep-frozen voyagers, Rip van Winkles awakened by robots when their destination is in sight. All such projects would be expensive and time-consuming, and although they may appear very unattractive to us, one can easily imagine cultures that would undertake them.

Travel to the stars *in a reasonable fraction of a human lifetime* is a much more difficult proposition, and it is this that arouses the ire of the physicists. For it necessarily involves speeds approaching that of light, and this appears to be beyond the bounds of engineering possibility. To see why, let us look at a few figures.

The nearest star—Alpha Centauri, a triple star system probably not suited for life—is about twenty-five trillion miles away. Since such a string of zeros is meaningless, the astronomers have invented the convenient unit of the light-year, or the distance that light travels in one year. (Let me emphasize that it is a unit of *distance;* many people seem to think it a measure of *time.*) Because light travels at 186,282 miles a second, simple arithmetic shows that a light-year is 5,880,000,000,000 miles—or six trillion, in round figures. So Alpha Centauri is 4.3 light-years away; there are about a dozen stars within ten light-years of us.

At the speed of light itself, therefore—assuming *no* time at all for starting and stopping, still less for sightseeing at the other end—the round trip to Alpha Centauri would take nearly nine years. This would not be impossible from the human point of view, though it would raise some pretty psychological problems. But it is, by many orders of magnitude, out of the question in terms of known engineering and known energy sources.

The best that our present rockets can do, and that with the greatest difficulty, is about nine miles a second. This is only one *twenty-thousandth* of the velocity of light—but there is worse to come, for energy increases with the *square* of velocity. To move a rocket twenty thousand times faster than the present limit we would need four hundred *million* times more energy. Even nuclear power comes nowhere near providing this. After some centuries of technical development, perhaps the most that we can hope for from hydrogen fusion is a tenth of the speed of light (say sixty million miles an hour!). This performance, which most physicists would consider highly optimistic, would just allow us to reach Alpha Centauri in a lifetime, for the one way voyage would last some fifty years.

Note that these calculations have nothing to do with any speed limit set by the theory of relativity; they are based purely on energy considerations. We simply do not know a source of energy sufficiently concentrated to drive a rocket anywhere near the speed of light.

However, it is always very dangerous to argue, on the basis of existing or even conceivable technology, that something can *never* be done. (See "The Hazards of Prophecy," *Profiles of the Future*.) In the past, those who have done so have almost invariably been proved wrong. What seemed to be insuperable obstacles have either been overcome, or simply bypassed by the development of new techniques. You cannot bridge the Golden Gate with wood—you have to wait until the steel age arrives; you cannot operate a TV system with ropes and pulleys—you have to wait until electronics comes along. If the rocket is inadequate for flight to the stars, which certainly appears to be the case, then we shall have to think of something better.

That there are several directions in which we may look is encouraging, but perhaps misleading; major breakthroughs are almost always quite unpredictable and occur in areas where no one would dream of finding them. (My favorite example: the greatest advance ever made in medicine resulted from a physicist's attempts to pass electricity through a vacuum tube. What had *that* to do with medicine? Only X rays.) In the case of interstellar flight, what we obviously need is a propulsion system which does *not* have to carry its source of energy with it, but can tap external supplies. The rocket is like a diesel or steam locomotive, limited in performance by the fuel it can carry. We require the equivalent of the electric locomotive—or perhaps the fuelless sailing ship.

Although electric fields, and swift but infinitely tenuous "winds," do exist in space, they are too feeble to be of any practical use. However, there are other cosmic forces and properties that we may some day utilize, as long ago we learned to use the moving airs and waters of this world for transportation. One of these forces, as was pointed out recently by Dr. Freeman J. Dyson, a highly imaginative mathematician at the Princeton Institute for Advanced Study, is gravity.

Dr. Dyson's conclusions are stimulating—and tantalizing. He suggests that the gravitational fields of certain double stars might be used, by sufficiently ingenious astronauts, to launch themselves out across interstellar space. Two stars, spinning rapidly round each other, could be used as a kind of cosmic slingshot, and during the period of acceleration the travelers would feel no force whatsoever. For a gravitational field acting upon a *freely* falling body produces no sense of weight: even if the astronauts were experiencing 10,000-g, and were thus increasing their speed at the enormous rate of 200,000 miles per hour *every second,* they would feel nothing at all as the stellar twins shot them off into space.

Unfortunately, we don't happen to have this particular type of double star (a white dwarf binary) in our immediate neighborhood. It is not even certain if such systems exist anywhere, but Dr. Dyson has an answer to this. To quote his words: "There may come a time in the remote future when engineering on *an astronomical scale* (my italics) will be both feasible and necessary." In other words if these "gravitational machines" do not exist in nature, they can be made.

Let us pause to give three hearty cheers to Dr. Dyson. His ideas may seem so farfetched that most people will regard them as extravagant fantasies, but when seen against the background of our incredible universe, they are entirely realistic. If we do not perform such feats in the millions of years that lie ahead, others will.

Another scheme for very high-speed cosmic flight depends on the fact that space is not entirely empty, but contains about ten atoms of hydrogen per cubic inch. For all ordinary purposes this is a perfect vacuum; however, a spaceship cruising at hundreds or thousands of miles a second would sweep up appreciable quantities of hydrogen. This leads to the daring concept of the "interstellar ramjet"—a device which would scoop up the hydrogen scattered between the stars, feed it into a fusion reactor, and spew out the resulting heated gases in a propulsive jet. It would, therefore, derive both its fuel and its working fluid from the space around it, and would thus have unlimited range.

Though the interstellar ramjet involves such fearsome technical problems that the first scientists to investigate the scheme rejected it out of hand, more recent studies have brought it back into favor. It is certainly centuries in the future, but it violates no fundamental principles. Even if it is never more than a theoretical concept, it is of great interest; for if *we* can think of slightly plausible ways of tapping the energies of space, we can be sure that our descendants will find much more practical ones.

And some day—perhaps by the use of beamed energy systems already glimpsed in the blinding light of the laser—we may learn to power our spacecraft from fixed ground stations. The analogy with the electric railroad would then be complete; spaceships need carry no fuel, as all their energy would be provided by planet-based installations which could be of unlimited size. This would again involve technologies far beyond our present horizon, but violating no basic laws. We need something like this to make space flight commercially practical even in the solar system; and what commerce needs, it eventually gets. If the rocket lasts as long as the steam engine, I shall be most surprised.

To sum up, then: Interstellar flight at speeds approaching that of light is *not* necessarily impossible, and those who have claimed that it is are being prematurely pessimistic. They may be right, but we shall not know for some centuries. Meanwhile, we will assume that they are wrong—and see just where this conclusion leads us.

In the old-fashioned Newtonian universe, which all scientists took for granted until the advent of Einstein, the situation was very straightforward. At the speed of light it would take you ten years to reach a star ten light-years away, and ten years to come home again. Total voyage time, twenty years. So if you were prepared to spend most of your life space-faring, you might roam thirty to forty light-years from Earth, and still return to your birthplace. If you wanted to do better than that, you had to travel faster than light. This would certainly be very difficult; but no one dreamed that it might be *impossible* for fundamental reasons concerned with the nature of the universe.

The special theory of relativity, published by Einstein in 1905, established a speed limit in space. There is nothing very mysterious about this, once it is understood that mass and energy are two sides of the same coin. If we accelerate an object, it gains energy by virtue of its speed. Therefore, it also gains mass—and the next time we try to increase its speed, we will find it correspondingly harder to accelerate. The effect is negligible at low velocities—that is, up to a few scores of millions of miles an hour!—which is why it was never detected in the past. For all ordinary purposes, the laws of motion laid down by Galileo and Newton still apply, as they always will.

But near the speed of light, the mass increase rises steeply. A law of diminishing returns sets in; though you may keep on pushing an object, its gain in speed is infinitesimal; all the additional energy goes into increasing its *mass*. There is nothing theoretical about this; few laws have been more thoroughly tested in practice, for billions of dollars' worth of engineering are now designed around it. The giant atom smashers—the Bevatrons, Cosmotrons, and so forth—are machines for accelerating nuclear particles to almost the speed of light. It requires thousands of tons of magnets and vacuum tubes to push the infinitely tiny electrons and protons up to these speeds, at which they may be hundreds of times heavier than when at rest. For this reason it has been suggested that nuclear accelerators should really be called "ponderators"; the increase in speed that they can produce at the end of their operating range is trivial, but the increase in mass is enormous.

At Berkeley and Brookhaven, and in myriads of high energy electronic devices (including the picture tube of your TV set) the Einstein equation is obeyed exactly. It predicts that even if we burned up the whole cosmos to accelerate a single electron, our infinitesimal "payload" would still fail to reach the speed of light. The solitary electron would have the mass of all the suns and galaxies that had been destroyed to propel it; but its speed would be only 99.999999999 . . . . —and not 100 per cent of that unattainable 186,282 miles per second. And what applies to one electron is true, a fortiori, to large-scale objects such as men and spaceships.

A sufficiently advanced civilization, by a prodigal expenditure of energy, might be able to drive its ships at 99 per cent of the speed of light. And since the remaining 1 per cent increase could never be attained anyway, there would seem to be no point in striving after it, merely to cut three days off every year of travel time.

However, matters are not as simple as this. The same equations which appear to limit us to journeys of a few dozen light-years in a single human lifetime also provide a loophole. Einstein once condemned a theory of which he did not approve with the words: "The good Lord is subtle, but He is never malicious." Nowhere is that subtlety more evident than in the laws that Einstein himself discovered.

Granted that we can never exceed the speed of light, it follows that a round trip from Earth to a star fifty light-years away can never take less than a hundred years. However—and this is something that no one had suspected before 1905—there is a profound ambiguity in our definition of time. Do we mean one hundred years to the crew of the spaceship, or to their friends waiting back on Earth?

For there *is* a distinction, and it took Einstein's genius to perceive it. The same equations that predict an increase of mass with velocity, also predict a stretching or dilatation of time, and according to precisely the same mathematical law. The discrepancy is negligible at low speeds, but becomes infinite at the speed of light. To a beam of light, time stands still; it can travel round the cosmos in one eternal instant.

The consequences of this are now well known; everyone has heard of the astronaut who sets out for the stars at almost the speed of light—and is still a young man when he returns to meet his aged twin brother, forty or fifty years later. Fantastic though this seems, it would actually occur if we could reach 99 per cent of the velocity of light, and within the narrow span of that last 1 per cent even more astonishing paradoxes would arise.

Here are some examples, given by the Harvard astronomer Carl Sagan in a paper with the splendid title "Direct Contact among Galactic Civilizations by Relativistic Interstellar Spaceflight." If a spaceship took off from Earth at a steady acceleration of one gravity (so that its occupants would feel their normal weight for the whole duration of the voyage), in five years by *ship time* it could reach a star ten light-years away. Yet it would not have exceeded the speed of light; to observers back on Earth, it would appear that the voyage had really lasted the full ten years. In effect, the clocks (and the people) on the spaceship would have run at half the speed of their counterparts on Earth—though they themselves would have noticed no change at all.

Ten years of *ship time* at 1-g acceleration would take the voyagers more than a hundred light-years, and thereafter the range goes up very steeply with time. Twenty years of cruising would bring them to the star clouds at the center of the Galaxy, some thirty thousand light-years from Earth. And in less than thirty years they would reach the Andromeda Nebula, more than a million light-years away! Of course, when the travelers returned home sixty years older, two million Earth years would have passed.

There is now no serious dispute about these conclusions; like the mass-increase laws, the time dilatation effect has been demonstrated experimentally. But perhaps I had better take a few minutes (Earth time) to dispose of an objection which is often raised to the so-called "clock paradox" by those who have a nodding acquaintance with relativity.

Because, they argue, Einstein stated that "all velocity is relative," it is just as legitimate to say that the spaceship is standing still and the Earth is moving. So the people on Earth should stay young, while the travelers age at the normal rate—which is obviously absurd.

Of course it is; but Einstein never said that all velocity is relative. That statement is true only of *uniform* velocities, and we are not dealing with these. The Earth is moving at a uniform speed, but the spaceship is steadily accelerating. So the two systems are not equivalent, and the paradox does not arise.

The theory of relativity, therefore, allows us to explore the universe without limit, by trading energy for time. Once again, it must be emphasized that the amounts of energy needed for such projects are gigantic, even by the standards of thermonuclear explosions. But they are not, in principle, beyond attainment or control; as Dr. Sagan concludes in his stimulating essay: "Allowing for a modicum of scientific and technological progress within the next few centuries, I believe that interstellar spaceflight at relativistic velocities to the farthest reaches of our Galaxy is a feasible objective for humanity."

There will, of course, be a price to pay, and it is not one that many of us would be prepared to face. Time would flow sluggishly in the speeding spaceship, but on Earth its progress would be inexorable. The voyagers would have cut themselves off forever from their friends and families, perhaps even from the culture that had launched them into space, if they returned hundreds or thousands of years in its future. For relativistic space flight is a kind of one-way time travel; though you can vary the rate at which the clock moves forward, you can never turn it back.

If Odysseus had sailed for Deneb, and not for Troy, we might expect him back at any moment, less grizzled than from his wanderings over the wine-dark sea. And how strange to think that, if ships from the galactic center visited our world in the remote past, there may *at this very moment* be a family of our Cro-Magnon ancestors on display in some celestial zoo. . . .

Most scientists who have convinced themselves that interstellar flight is possible believe that such visits must have occurred, perhaps many times in the long history of Earth. The astronomer Thomas Gold has even suggested that terrestrial life arose from the garbage dumped by one of these early expeditions. I should love to see somebody found a religion on this inspiring belief; but odder faiths have flourished in the past.

There are many who will be profoundly dissatisfied with these conclusions, and will feel aggrieved because we can never race back and forth across the universe as we now do over the face of this Earth. They may even doubt the eternal validity of the Einstein equations, though these have stood unchallenged for half a century, and are now backed by the awesome authority of the mushroom cloud.

After all, many other apparent limits have proved to be no more than temporary road blocks. Less than twenty years ago, we were worrying about the sound barrier; tomorrow, grandmothers will be cruising at Mach 3. Will the "light barrier" go the same way?

I am afraid I cannot offer much hope. If you have followed me so far, you will have realized the utterly fundamental nature of this barrier. And it is no good asking *why* we cannot travel faster than light, and *why* time dilatation occurs; our universe is simply built that way. Anyone who doesn't like it can go somewhere else.

Perhaps that last sentence offers the one faint chance of beating Einstein. If other universes—other space-time continua—do exist, light may propagate in them at higher speeds than our familiar 186,282 miles a second. We may be able to get to the Andromeda Nebula *and back again* in a few years of Earth time, by taking a spatial detour through another dimension. But this is pure fantasy, with no scientific basis; so is the suggestion that we might be able to tap the so-called *psi* or paranormal forces which some students claim to have detected. If cosmic teleportation is practical, the current paucity of visitors becomes even more difficult to explain. Unless we are under quarantine (a highly plausible assumption), it really looks as if interstellar travel is expensive, time-consuming and, therefore, infrequent.

We had better cooperate with the inevitable, and, after all, we have no great reason to complain. This planetary system will keep us busy for quite a while, and beyond that, there are some four hundred stars of roughly solar type within a hundred light-years.

So even if we cannot exorcise the ghost of Einstein (and what *were* those dying words of his, lost forever because his nurse understood no German?), we have a prospect before us that will daunt whole armies of biologists and historians. Columbus is not yet five hundred years in the past; but before another five centuries have gone, we may have complete records of a hundred civilizations, most of them far older than our own.

We may well be grateful, then, that our sphere of knowledge cannot expand more swiftly than light. That speed limit may be the only thing that can save us, when the *real* space age dawns, from being utterly overwhelmed by the richness and complexity of our many-splendored universe.

The Light of Common Day

No man has ever seen the Sun, or ever will. What we call "sunlight" is only a narrow span of the entire solar spectrum—the immensely broad band of vibrations which the Sun, our nearest star, pours into space. All the colors visible to the eye, from warm red to deepest violet, lie within a single octave of this band—for the waves of violet light have twice the frequency, or "pitch" if we think in musical terms, of red. On either side of this narrow zone are ranged octave after octave of radiations to which we are totally blind.

The musical analogy is a useful one. Think of one octave on the piano—less than the span of the average hand. Imagine that you were deaf to all notes outside this range; how much, then, could you appreciate of a full orchestral score when everything from contrabassoon to piccolo is going full blast? Obviously you could get only the faintest idea of the composer's intentions. In the same way, by eye alone we can obtain only a grossly restricted conception of the true "color" of the world around us.

However, let us not exaggerate our visual handicap. Though visible light is merely a single octave of the Sun's radiation, this octave contains most of the power; the higher and lower frequencies are relatively feeble. It is, of course, no coincidence that our eyes are adapted to the most intense band of sunlight; if that band had been somewhere else in the spectrum, as is the case with other stars, evolution would have given us eyes appropriately tuned.

Nevertheless, the Sun's invisible rays are extremely important, and affect our lives in a manner undreamed of only a few years ago. Some of them, indeed, may control our destinies—and even, as we shall see in a moment, our very existence.

The visible spectrum is, quite arbitrarily, divided up into seven primary colors—the famous sequence, red, orange, yellow, green, blue, indigo, violet, if we start from the longest waves and work down to the shortest. Seven main colors in the one octave; but the complete band of solar radiations covers at least thirty octaves, or a total frequency range of ten thousand-million to one. If we could see the whole of it, therefore, we might expect to discern more than *two hundred* colors as distinct from each other as orange is from yellow, or green is from blue.

Starting with the Sun's visible rays, let us explore outward in each direction and see (though that word is hardly applicable) what we can discover. On the long-wave side we come first to the infrared rays, which can be perceived by our skin but not by our eyes. Infrared rays are heat radiation; go out of doors on a summer's day, and you can tell where the Sun is even though your eyes may be tightly closed.

Thanks to special photographic films, we have all had glimpses of the world of infrared. It is an easily recognizable world, though tone values are strangely distorted. Sky and water are black, leaves and grass dazzling white as if covered with snow. It is a world of clear, far horizons, for infrared rays slice through the normal haze of distance—hence their great value in aerial photography.

The further we go down into the infrared, the stranger are the sights we encounter and the harder it becomes to relate them to the world of our normal senses. It is only very recently (partly under the spur of guided missile development) that we have invented sensing devices that can operate in the far infrared. They see the world of heat; they can "look" at a man wearing a brilliantly colored shirt and smoking a cigarette—and see only the glowing tip. They can also look down on a landscape hidden in the darkness of night, and see all the sources of heat from factories, automobiles, taxiing aircraft. Hours after a jet has taken off, they can still read its signature on the warm runway.

Some animals have developed an infrared sense, to enable them to hunt at night. There is a snake which has two small pits near its nostrils, each holding a directional infrared detector. These allow it to "home" upon small, warm animals like mice, and to strike at them even in complete darkness. Only in the last decade have our guided missiles learned the same trick.

Below the infrared, for several octaves, is a no-man's-land of radiation about which very little is known. It is hard to generate or to detect waves in this region, and until recently few scientists paid it much attention. But as we press on to more familiar territory, first we encounter the inch-long waves of radar, then the yard-long one of the shortwave bands, then the hundred-yard waves of the broadcast band.

The existence of all these radiations was quite unknown a century ago; today, of course, they are among the most important tools of our civilization. It is a bare twenty years since we discovered that the Sun also produces them, on a scale we cannot hope to match with our puny transmitters.

The Sun's radio output differs profoundly from its visible light, and the difference is not merely one of greater length. Visible sunlight is practically constant in intensity; if there are any fluctuations, they are too slight to be detected. Not only has the Sun shone with unvarying brightness throughout the whole span of human history, but we would probably notice no difference if we could see it through the eyes of one of the great reptiles.

But if you saw only the "radio" Sun, you would never guess that it was the same object. Most of the time it is very dim—much dimmer, in fact, than many other celestial bodies. To the eye able to see only by radio waves, there would be little difference between day and night; the rising of the Sun would be a minor and inconspicuous event.

From time to time, however, the radio Sun explodes into nova brightness. It may, *within seconds,* flare up to a hundred, a thousand, or even a million times its normal brilliance. These colossal outbursts of radio energy do not come from the Sun as a whole, but from small localized areas of the solar disc, often associated with sunspots.

This is one excellent reason why no animals have ever developed radio senses. Most of the time, such a sense would be useless, because the radio landscape would be completely dark—there would be no source of illumination.

In any event, "radio eyes" would pose some major biological problems, because radio waves are millions of times larger than normal eyes, if they were to have the same definition. Even a radio eye which showed the world as fuzzily as a badly out-of-focus TV picture would have to be hundreds of yards in diameter; the gigantic antennas of our radar systems and radio telescopes dramatize the problem involved. If creatures with radio senses do exist anywhere in the universe, they must be far larger than whales, and can, therefore, only be inhabitants of gravity-free space.

Meanwhile, back on Earth, let us consider the other end of the spectrum—the rays shorter than visible light. As the blue deepens into indigo and then violet, the human eye soon fails to respond. But there is still "light" present in solar radiation: the ultraviolet. As in the case of the infrared, our skins can react to it, often painfully; for ultraviolet rays are the cause of sunburn.

And here is a very strange and little-known fact. Though I have just stated that our eyes do not respond to ultraviolet, the actual situation is a good deal more complicated. (In nature, it usually is.) The sensitive screen at the back of the eye—the retina, which is the precise equivalent of the film in a camera—*does* react strongly to ultraviolet. If it were the only factor involved, we could see by the invisible ultraviolet rays.

Then why don't we? For a purely technical reason. Though the eye is an evolutionary marvel, it is a very poor piece of optics. To enable it to work properly over the whole range of colors, a good camera has to have four, six or even more lenses, made of different types of glass and assembled with great care into a single unit. The eye has only one lens, and it already has trouble coping with the two-to-one range of wavelengths in the visible spectrum. You can prove this by looking at a bright red object on a bright blue background. They won't both be in perfect focus; when you look at one, the other will appear slightly fuzzy.

Objects would be even fuzzier if we could see by ultraviolet as well as by visible light, so the eye deals with this insoluble problem by eliminating it. There is a filter in the front of the eye which blocks the ultraviolet, preventing it from reaching the retina. The haze filter which photographers often employ when using color film does exactly the same job, and for a somewhat similar reason.

The eye's filter is the lens itself—and here at last is

the punch line of this rather long-winded narrative. If you are ever unlucky enough to lose your natural lenses (say through a cataract operation) and have them replaced by artificial lenses of clear glass, you will be able to see quite well in the ultraviolet. Indeed, with a source of ultraviolet illumination, like the so-called "black light" lamps, you will be able to see perfectly in what is, to the normal person, complete darkness! I hereby donate this valuable information to the C.I.A., James Bond, or anyone else who is interested.

Normal sunlight, as you can discover during a day at the beach, contains plenty of ultraviolet. It all lies, however, in a narrow band—the single octave just above the visible spectrum in frequency. As we move beyond this to still higher frequencies, the scene suddenly dims and darkens. A being able to see *only* in the far ultraviolet would be in a very unfortunate position. To him, it would always be night, whether or not the sun was above the horizon.

What has happened? Doesn't the Sun radiate in the far ultraviolet? Certainly it does, but this radiation is all blocked by the atmosphere, miles above our head. In the far ultraviolet, a few inches of ordinary air are as opaque as a sheet of metal.

Only with the development of rocket-borne instruments has it become possible to study this unknown region of the solar spectrum—a region, incidentally, which contains vital information about the Sun. If you started off from ground level on a bright, sunny day, this is what you would see:

At first, you would be in utter darkness, even though you were looking straight at the Sun. Then, about twenty miles up, you would notice a slow brightening, as you climbed through the opaque fog of the atmosphere. Beyond this, between twenty and thirty miles high, the ultraviolet Sun would break through in its awful glory.

I use that word "awful" with deliberate intent. These rays can kill, and swiftly. They do not bother astronauts, because they can be easily filtered out by special glass. But if they reached the surface of the Earth—if they were not blocked by the upper atmosphere—most existing forms of life would be wiped out.

If you regard the existence of this invisible ultraviolet umbrella as in any way providential, you are confusing cause and effect. The screen was not put in the atmosphere to protect terrestrial life. It was put there by life itself, hundreds of millions of years before man appeared on Earth.

The Sun's raw ultraviolet rays, in all probability, *did* reach the surface of the primeval Earth; the earliest forms of life were adapted to it, perhaps even thrived upon it. In those days, there was no oxygen in the atmosphere; it is a by-product of plant life, and over geological aeons its amount slowly increased, until at last those oxygen-burning creatures called animals had a chance to thrive.

That filter in the sky is made of oxygen—or, rather, the grouping of three oxygen atoms known as ozone. Not until Earth's protective ozone layer was formed, and the short ultraviolet rays were blocked twenty miles up, did the present types of terrestrial life evolve. If there had been no ozone layer, they would doubtless have evolved into different forms. Perhaps we might still be here, but our skins would be very, very black.

Life on Mars must face this problem, for that planet has no oxygen in its atmosphere, and, therefore, no ozone layer. The far ultraviolet rays reach the Martian surface unhindered, and must profoundly affect all living matter there. It has been suggested that these rays are responsible for the color changes which astronomers have observed on the planet. Whether or not this is true, we can predict that one of the occupational hazards of Martian explorers will be severe sunburn.

Just as ultraviolet lies beyond the violet, so still shorter rays lie beyond it. These are X rays, which are roughly a thousand times shorter than visible light. Like the ultraviolet, these even more dangerous rays are blocked by the atmosphere; few of them come to within a hundred miles of Earth, and they have been detected by rocket instruments only during the last few years. The solar X rays are quite feeble—only a millionth of the intensity of visible light—but their importance is much greater than this figure would indicate. We know now that blasts of X rays from the Sun, impinging upon the upper atmosphere, can produce violent changes in radio communications, even to the extent of complete blackouts.

Men have lost their lives because the Sun has disrupted radio; nations are equally vulnerable, in this age of the ICBM.

You will recall that though the Sun shines with remarkable steadiness in the visible spectrum, it flares and sparkles furiously on the long (radio) waves. Exactly the same thing happens with its X-ray emission, even though these waves are a billion times shorter. Moreover, both the Sun's radio waves and its X rays appear to come from the same localized areas of the solar surface—disturbed regions in the neighborhood of sunspots, where clouds of incandescent gas larger than the Earth erupt into space at hundreds of miles a second.

For reasons not yet understood (there is not much about the Sun that we do *thoroughly* understand) solar activity rises and falls in an eleven-year cycle. The Sun was most active around 1957, which is why that date was chosen for the International Geophysical Year. Now it is heading for a minimum, and to take advantage of this, scientists are making arrangements for a little IGY called the "Year of the Quiet Sun." It is rather unfortunate that the Sun will be coming back to the boil at exactly the time that the first major space expeditions are being planned—say around 1968. The astronauts may run into some heavy weather, for by then the Sun will be shooting out not only vast quantities of ultraviolet, X rays and radio waves, but other radiations which cannot be so easily blocked.

We see, then, how complicated and how variable sunlight is, if we use that word in the widest sense to describe all the waves emitted by the Sun. Nevertheless, when we accept the evidence of our unaided eyes and describe the Sun as a yellow star, we have summed up the most important single fact about it—*at this moment in time*. It appears probable, however, that sunlight will be the color we know for only a negligibly small part of the Sun's history.

For stars, like individuals, age and change. As we look out into space, we see around us stars at all stages of evolution. There are faint blood-red dwarfs so cool that their surface temperature is a mere 4,000 degrees Fahrenheit; there are searing ghosts blazing at 100,000 degrees, and almost too hot to be seen, for the greater part of their radiation is in the invisible ultraviolet. Obviously, the "daylight" produced by any star depends upon its temperature; today (and for ages past, as for ages to come) our Sun is at about 10,000 degrees F, and this means that most of its light is concentrated in the yellow band of the spectrum, falling slowly in intensity toward both the longer and the shorter waves.

That yellow "hump" will shift as the Sun evolves, and the light of day will change accordingly. It is natural to assume that as the Sun grows older, and uses up its hydrogen fuel—which it is now doing at the spanking rate of half a billion tons *a second*—it will become steadily colder and redder.

But the evolution of a star is a highly complex matter, involving chains of interlocking nuclear reactions. According to one theory, the Sun is still growing hotter, and will continue to do so for several billion years. Probably life will be able to adapt itself to these changes, unless they occur catastrophically, as would be the case if the Sun exploded into a nova. In any event, whatever the vicissitudes of the next five or ten billion years, at long last the Sun will settle down to the white dwarf stage.

It will be a tiny thing, not much bigger than the Earth, and, therefore, too small to show a disc to the naked eye. At first, it will be hotter than it is today, but because of its minute size it will radiate very little heat to its surviving planets. The daylight of that distant age will be as cold as moonlight, but much bluer, and the temperature of Earth will have fallen to 300 degrees below zero. If you think of mercury lamps on a freezing winter night, you have a faint mental picture of high noon in the year 7,000,000,000 A.D.

Yet that does not mean that life—even life as we know it today—will be impossible in the solar system; it will simply have to move in toward the shrunken Sun. The construction of artificial planets would be child's play to the intelligences we can expect at this date; indeed, it will be child's play to us in a few hundred years' time.

Around the year 10,000,000,000 the dwarf Sun will have cooled back to its present temperature, and hence to the yellow color that we know today. From a body that was sufficiently close to it—say only a million miles away—it would look exactly like our present Sun, and would give just as much heat. There would be no way of telling, by eye alone, that it was actually a hundred times smaller, and a hundred times closer.

So matters may continue for another five billion years; but at last the inevitable will happen. Very slowly, the Sun will begin to cool, dropping from yellow down to red. Perhaps by the year 15,000,000,000 it will become a red dwarf, with a surface temperature of a mere 4,000 degrees. It will be nearing the end of the evolutionary track, but reports of its death will be greatly exaggerated. For now comes one of the most remarkable, and certainly least appreciated, results of modern astrophysical theories.

When the Sun shrinks to a dull red dwarf, it will not be dying. It will just be starting to live—*and everything that has gone before will be merely a fleeting prelude to its real history*.

For a red dwarf, because it is so small and so cool, loses energy at such an incredibly slow rate that it can stay in business for *thousands* of times longer than a normal sized white or yellow star. We must no longer talk in billions, but of trillions of years if we are to measure its life span. Such figures are, of course, inconceivable (for that matter, who can think of a thousand years?). But we can nevertheless put them into their right perspective if we relate the life of a star to the life of a man.

On this scale, the Sun is but a week old. Its flaming youth will continue for another month; then it will settle down to a sedate adult existence which may last at least eighty years.

Life has existed on this planet for two or three days of the week that has passed; the whole of human history lies within the last second, and there are eighty years to come.

In the wonderful closing pages of *The Time Machine,* the young H. G. Wells described the world of the far future, with a blood-red Sun hanging over a freezing sea. It is a somber picture that chills the blood, but our reaction to it is wholly irrelevant and misleading. For we are creatures of the dawn, with eyes and senses adapted to the hot light of today's primeval Sun. Though we should miss beyond measure the blues and greens and violets which are the fading afterglow of Creation, they are all doomed to pass with the brief billion-year infancy of the stars.

But the eyes that will look upon that all-but-eternal crimson twilight will respond to the colors that we cannot see, because evolution will have moved their sensitivity away from the yellow, somewhere out beyond the visible red. The world of rainbow-hued heat they see will be as rich and colorful as ours—and as beautiful; for a melody is not lost if it is merely transposed an octave down into the bass.

So now we know that Shelley, who was right in so many things, was wrong when he wrote:

> Life, like a dome of many-colored glass,
> stains the white radiance of eternity.

For the radiance of eternity is not white: It is infrared.

Last night I saw a great liner—the 42,000 ton *Oriana* —steam out of Colombo harbor on her maiden voyage, with lights ablaze and rockets raking the sky. As she dwindled round the curve of the world, heading toward the equator and the southern stars, my imagination sailed beyond her, into the past and into the future. I thought of all that the single word "ship" has meant to man, during the centuries that he has used the sea as a highway and a battlefield. And I thought of the ships that he would soon be launching into the shoreless seas of space.

It is a long line, and a splendid one, from the legendary *Argo* to the gigantic nuclear-powered aircraft carrier of today. One could write a history of the world whose chapter headings were the names of ships; such a roll call would include *Santa Maria, Golden Hind, Endeavour, Victory, Beagle, Great Eastern, Monitor, Nautilus.* All are associated with key events in the story of man, his discovery of his planet —and of himself. It would be hard to say whether the voyage of Charles Darwin aboard the *Beagle,* or of Columbus aboard the *Santa Maria,* had a greater ultimate impact upon the world.

Of all the creations of man's brains and hands, his ships are perhaps the noblest, for them at least, the taunting words that Shaw put into the mouth of the Devil are not always true. "Man's industrial machinery," says Mephistopheles to Don Juan in *Man and Super-*

man, "is the product of his greed and sloth; his heart is in his weapons." Though one hesitates, on general principles, to contradict such an authority, it is a gross oversimplification to maintain that only greed inspired the splendid tea clippers of the nineteenth century, or the Atlantic liners of today. Of course they were built to make money, but men put their hearts as well as their bank balances into them; and often they lost both. "It was the care and effort that went into her—she was a beautiful, wonderful ship," lamented a survivor of the *Titanic.* Those words have been echoed in many times, and in many languages.

It is easy to see why ships have such an appeal to the human spirit. With their mobility, they symbolize freedom and adventure. They are man's tribute to the mother he loves and hates—the sea. Nothing else he has ever built combines so perfectly the attributes of power and beauty; to find any comparison, one must go back to the Middle Ages. The great ocean liners are the cathedrals of modern man.

Now he is about to build new shrines to later gods, for the ships of the future will be ships of space. They will embody power beyond all the dreams of the past, but beauty, it now appears, must remain on Earth; it will not survive the passage through the atmosphere.

Those last words can be taken in an almost literal sense; if we look once more at the sea, we will understand why. There is a natural—if you like, a "right"— shape for everything that moves on or under the water; it is determined by the laws of hydrodynamics. Even if we had never built ships or submarines, we could have learned this fact merely by looking at the fish. Over the centuries, without conscious thought, we have accepted nature's maxims; they are now part of our aesthetic makeup. Though styles in ships may change for technical reasons (this century has seen four funnels dwindle to two and then to none) there is a wide measure of agreement between seafarers and landsmen as to what constitutes a good-looking ship. It is no surprise to learn that the *Great Eastern,* considered by many to be the most beautiful steamship ever built, was designed by the only engineer-artist to come within hailing distance of Leonardo during the last four hundred years.

In the newer field of aviation, the aesthetic content is even more obvious; look at the Spitfire or the Comet. Until recently, beauty and performance have always gone hand-in-hand in aeronautics, but even here we have come to the parting of the ways.

For now we are entering a region where all the centuries of human experience and all the aeons of natural selection, can give us no guidance at all. As we sail out into space, we must be prepared to forget all our terrestrial ideas and preconceptions. The familiar shapes of fish and birds and falling raindrops have no significance here, and must not be allowed to influence our thinking. When they do, the results can be dangerously misleading.

A perfect example of this is provided by the re-entry vehicles—ICBM nose cones and Apollo capsules—which have evolved during the past decade. It seemed obvious that a body returning to the atmosphere at 15,000 miles an hour must have the ultimate in needle-nosed streamlining, and the earlier vehicles were so designed. But then it was discovered, after several years and a few hundred million dollars, that all our intuitive ideas in this region were completely wrong. Today's re-entry vehicles are round-nosed or even blunt-ended structures, about as well streamlined as milk bottles.

A line-up of the space vehicles developed over the last twenty years tells the same story. The ancestor of them all, the V.2 rocket (designed 1938–1942) is the only one that looks right. It is nicely streamlined, with fins in the appropriate places, just like the spaceships in the old science fiction magazines.

But look at its successors! The fins have gone; nowadays rockets do all the steering. The only concession to streamlining is a rounded nose, often jettisoned as soon as the lower atmosphere is safely past. As for the body itself, it is now almost invariably a straight-sided cylinder, or a series of them. There are no sweeping lines to delight the eye; the giant rocket of today is merely a conglomeration of storage tanks in a hurry.

Aesthetically, the situation is even worse where deep space vehicles are concerned. By this term is meant probes, satellites and ultimately manned spacecraft designed to spend all their careers outside the atmosphere, far beyond the last traces of air resistance.

The space vehicles now being built are collections of antennas, instruments, batteries, jet nozzles, TV cameras, gas cylinders, and boxes of electronics, held together by frameworks of light struts. Insofar as they have any natural counterparts on Earth, we might find them among the articulated nightmares of the insect world, or the even more *outré* beings which float across the field of the microscope when a drop of plankton-filled sea water is examined. They can arouse curiosity, amazement, disquiet or even dismay—but hardly affection. Many of them bear an uncanny resemblance to the products of the "Junk Heap" school of sculpture, which finds its raw materials in the scrap piles of our mechanical age. (See, for example, César's suggestively titled *Galactic Insect* at the New York Museum of Modern Art. It is made from iron rings, metal mesh and miscellaneous bars and rods welded together, and no one would care to meet it on a dark night.)

The man-carrying vehicles of the next decade—the first true spaceships—will be larger than today's Mariners and Rangers, but no more beautiful. They will consist of pressurized spheres or cylinders housing the crew, large tanks holding fuel for the rockets, panels of solar cells spread like sails to collect electricity from the sun, and many other fixed or movable appendages. It will be impossible to tell which way they are going—for in space, of course, this statement has no meaning. Out there, one direction is as good as any other.

These odd, bristling structures will carry the commerce of the future between the planets, shuttling from orbit to orbit but never making a landing on any world. Because they will be traveling through unresisting vacuum and will be weightless, like all bodies orbiting freely through space, they will be free from the structural limitations that affect vehicles and buildings on Earth. They could be of any shape or size that their particular function required, and could be made of materials almost as flimsy as Chinese lanterns. Though the first generation space vehicles are small enough to be lifted beyond the atmosphere in a single rocket (Saturn V can put a hundred tons in orbit) the later and much larger ones will be assembled in space, from materials ferried up by successive flights. There is no limit to the size which might be ultimately attained; the space liners of the far future may completely dwarf the greatest ocean giants of today.

Some of them—particularly those specializing in passengers rather than freight—will be shaped like wheels or drums and will slowly revolve around their axis. The riders on such cosmic carousels will thus be given an artificial but convincing sensation of gravity, as centrifugal force converts the outer wall of the spinning ship into a floor. A space liner whirling along its orbit will be an intriguing sight, but hardly an elegant one.

It is obvious that ships of space are going to have very little in common with ships of the sea, but they will have even less in common with aircraft. If I wished to add fuel to an inflammable argument, I could make a very good case for putting the Navy rather than the Air Force in charge of space operations. How many bomber missions last for weeks, months or even years? How many airmen have ever had to land upon, explore and survey wholly unknown and possibly hostile territory? The only parallels we have with the space voyages of the future come from the great maritime expeditions of the past. The explorers of tomorrow will be nearer to Magellan and Cook than to Lindbergh or any of the other pioneers who blazed a trail across Earth's skies during the brief heroic age of aviation.

Perhaps the most striking respect in which ships of space will differ from their terrestrial precursors is this: Barring very unusual accidents, they will be immortal.

Marine craft have only a limited life; corrosion, the wear and tear of the sea, technological obsolescence —these ensure that very few ships outlive the men who build them. And although shipwreck is now somewhat unusual, it still happens more often than is generally realized. If I took a bet that a ship is sinking at this very minute in some ocean of the world, I should probably win.

A spaceship can never sink, and though it may be disabled, it can seldom be destroyed. Though many things could go wrong in so complex a device as a spacecraft, the possibilities of *complete* destruction are very limited. The most obvious (and perhaps most likely) cause of disaster would be a violent internal explosion; fire in space could be an even greater peril than fire at sea.

Even a very large meteor could not cause much physical damage to a spaceship, though it might put it out of action by drilling a hole through essential machinery. Meteors more than a few inches in diameter are so excessively rare that we may never encounter them in centuries of space-faring. If a spaceship ever collides with another heavenly body, it will be through pilot error or engine failure during take-off or approach; it is almost impossible to hit another planet unintentionally. If you look up at the night sky, and then remind yourself that only three or four of those points of light lie within twenty trillion miles of Earth, you will see what chance a spaceship has of accidentally running aground. No Robinson Crusoe of the future will ever be cast upon a desert asteroid.

Even if a spaceship were disabled, and became an inert mass of metal, carrying its crew through the solar system in a celestial tomb, it would not be lost. It would become another planet of the Sun, as has already happened with our first deep-space probes, Lunik

I, Pioneer V, and their successors. Though they are now scores of millions of miles away from the Earth, we know their approximate positions, and a radar search may one day lead to their recovery and display in some museum of astronautics. After a hundred or a thousand years in space, they will be as good as new, apart from some superficial pitting by meteor dust.

The position of future space wrecks will be known to a fantastic degree of accuracy. We can be quite sure that the departure from Earth of any large man-carrying vessel will be watched by precision radars; even now, there is tracking equipment at Cape Kennedy with an error of less than one foot *in a thousand miles*. And once a space vehicle has been launched into its orbit, there are no winds or currents to push it off course. The only forces acting upon it are the gravitational tugs of the Sun and planets, which are known with very great accuracy. If a spaceship's position and velocity are pinpointed just once, its future movements can be calculated until the end of time, by the methods that the astronomers worked out long ago to predict the motions of the planets.

There seems only one way in which a spaceship could become lost, in the maritime sense of "lost without trace." That would be if, through some error or accident, it used its rockets to make a major change of speed that was neither reported back to Earth, nor observed by any tracking station. Then, instead of traveling along its assumed orbit, the spaceship would head off on a completely different one, and might soon be so many millions of miles away from the calculated position that it could never again be located. Space is so unimaginably huge that searching for a needle in a haystack would be a childishly simple operation, compared with hunting down a rocket that had gone off on a wild trajectory.

Even if we knew the exact position of a derelict spaceship, the information would often be of little value; the cost of going after it might be so great that it would be, for all practical purposes, inaccessible. As every object in the solar system is swinging round the Sun at widely varying distances and in periods that range from two and a half months for Mercury to two and a half centuries for Pluto, a close encounter between a spaceship and a planet could occur only at enormous intervals. A ship that shot away from the Sun on a long orbit like that of a comet might not return to the neighborhood of the Earth for a million years. We would know exactly where it was for the rest of eternity, but could do nothing about it. The quantity of rocket fuel needed to go after it *and come back again* might be so great that no ship could carry it.

Space dramas and tragedies which have no precise counterpart on Earth will, without doubt, result one day from these facts of celestial mechanics. If anything went wrong with its propulsion system, even if all its other equipment was functioning perfectly, a spaceship would be doomed. There might be enough air and food aboard to keep its passengers alive for months, but that would only prolong the agony, for the chances of a rescue operation in deep space are almost vanishingly small.

It is not merely that space is so enormous, but that everything is moving so quickly. If we travel from one place to another on this planet, we know that both our destination and our point of origin are fixed; whatever *we* do, they will always be in the same place. When a jet liner leaves London for New York, her captain can rest assured that Kennedy Airport will not drift several thousand miles westward during the course of the flight—or be moving at six hundred miles an hour when he tries to land.

But this is a fair analogy of what would happen on every interplanetary journey. To make a safe arrival, you have to be in the right place, at the right time, and *moving at the right speed*. It is this last requirement that will make "space-rescue" operations so extraordinarily difficult that they can seldom be attempted.

All spacecraft, for as far ahead as we can imagine, will have to operate within very tight fuel budgets.

One of the hardest things for the layman to realize is that two spaceships might be in the same neighborhood—might, indeed, pass within yards of each other —yet be as far apart, for all practical purposes, as if they were on opposite sides of the solar system. They could exchange signals, but they could make no physical contact unless their speeds had been exactly matched. A would-be rescue ship could race helplessly past a space wreck without being able to assist it in any way, if its fuel supply was insufficient both to match speed *and,* ultimately, to depart again on an orbit that would take it back to safety. If it could only carry out the contact operation and not the departure, the result would merely be two derelicts instead of one.

What heartbreaking tragedies of the future will flow from this! How will those in charge of space operations be able to explain to the public that the crews and passengers crying on the radio for help are already dead men, their fates sealed by inexorable mathematics though they may still have months to live? If the available ships could not carry enough fuel for a rescue mission, no amount of skill and courage would help; the decision of the computers would be final and beyond argument.

It would be quite possible for a runaway spaceship to escape from our Sun altogether, and go heading out into the interstellar night. Such accidents may have happened elsewhere, countless times in the long history of the universe; how many ships of unknown registry are drifting between the stars, and who will be the first men to board such strangers? All the treasures of all the wrecks of Earth may be as nothing, compared with the argosies of scientific knowledge that may be floating in the deeps of space, perfectly preserved from the ravages of time. Perhaps the only record of many races, which perished millions or billions of years ago, may be in the ships and the cargoes they have lost. As we today are recovering the drowned legends of our own past from the bed of the Mediterranean, so archeologists of the future may reconstruct the histories of other beings from the flotsam and jetsam of space.

A few years ago, such thoughts were the exclusive province of science fiction; today, astronomers and physicists are discussing them in learned letters to *Nature*. Almost every scientist now accepts as possible, if not probable, the existence of intelligent life elsewhere in the universe; once that is granted, it is ridiculous to assume that all the vehicles in space will be put there by man.

A leading Soviet mathematician has argued forcibly that one alien space vehicle has already been discovered. According to Dr. I. S. Shklovsky, the inner satellite of Mars (and perhaps the outer one as well) may be of artificial origin.

Dr. Shklovsky's reasoning is ingenious yet (unlike many mathematical arguments) beautifully simple. Though he may not have convinced anyone, he has undoubtedly made many astronomers do some hard thinking, and no one has yet given a better explanation of the peculiar facts he has marshaled.

Phobos is an extremely small body, as natural moons go; its diameter is roughly ten miles. It revolves around Mars in about eight hours—less than a third of the time Mars itself takes to turn on its axis. No other moon goes round its planet in less than that planet's "day," but all our artificial satellites do so, this side of *Early Bird*.

However, the really surprising thing about Phobos is that it is slowly falling down to Mars. This can only mean that minute traces of air resistance, even at its height of thirty-five hundred miles, are robbing it of its energy—as, once again, is happening with our own satellites, many of which have already burned up in the atmosphere. It will be a good many million years before this happens to Phobos, but its eventual fate is beyond dispute.

Now, the rate at which a satellite—be it natural or artificial—is dragged down by air resistance depends upon its density. A satellite made of heavy, compact materials will stay up very much longer than a light, hollow one. By observing the rate at which a satellite descends, it is possible to calculate its density, and it turns out, according to Dr. Shklovsky, that Phobos is a thousand times lighter than water.

If his figure is correct, there is only one conclusion to be drawn: Phobos must be hollow, little more than a gigantic shell or balloon. It is practically impossible to imagine a natural explanation for such a phenomenon, and the analogy with our own Echo balloon satellites is striking. True, a ten mile diameter structure is an impressive thought, but in gravityless space, as I have already remarked, size is no object. Within three years of entering space, we have launched hundred foot satellites; what may we not be doing a century from now?

Dr. Shklovsky's stimulating theory appeals to me because some ten years ago I made an identical suggestion concerning the innermost moon of Jupiter. In a story called "Jupiter V" I pointed out certain peculiarities of this satellite and developed the idea that it was a giant spacecraft which, ages ago, had entered the solar system and then been "parked" in orbit round Jupiter while its occupants went off in more conveniently sized vehicles to colonize the planets.

If the interstellar gulfs have ever been crossed by living creatures, they may well have needed spacecraft so large that they were virtually self-contained artificial worlds, within which generation after generation of travelers could be born and die, never knowing any environment other than that of the ship. The histories of such space arks would make all the sagas of Earth seem puny indeed; even now, perhaps, many of them are being sung by strange poets in tongues that we shall never know.

But before intelligent creatures set off on such immense journeys, they would wish to know a great deal about their destination. Just as we are doing today, but on a far larger scale, they would launch robot surveyors ahead of them. Perhaps they could never do more than this; the chasms between the stars may be too wide ever for life to cross them. But they would be no obstacle to patient, immortal machines which could sleep for centuries and then awake to carry out the instructions of their long-dead builders.

Before the first decade of the space age draws to a close, we ourselves will have the power to launch small payloads of instruments to Alpha Centauri, four and a quarter light-years away. The one-way trip for a vehicle which left the solar system at the speed of today's space probes would last over a hundred thousand years, but foreseeable advances would bring the time down to a few centuries. At that stage, we may well consider sending automatic probes to look at our stellar neighbors, and to report back a millennium hence. Unless we accept the almost incredible hypothesis that *we* are the only intelligent, space-minded beings in the entire Galaxy of a hundred thousand million suns, we must assume that innumerable robot explorers are shuttling back and forth amid the stars. After all, we ourselves are now launching one a week, and who would have credited *that* five years ago?

Men have been building ships for at least six thousand years; the earliest known wreck is dated 1400 B.C.[1] Yet there may be ships far older than this plying between the stars—not as wandering derelicts, but as fully operational craft, going about missions that were conceived and planned when our ancestors were chipping their first flints. These Flying Dutchmen of space, obeying the wills of creatures who could never hope to know whether their efforts would meet with failure or success, may be waiting to greet us when we embark upon our own cosmic voyages.

If such automatic explorers have entered our solar system, they will have homed at once upon Earth, now that it is pouring out so many millions of watts of radio energy. The fact that we have, as yet, received no intelligent replies from space may be because we have

---

[1] When I wrote this I little suspected that one day its discoverer, Peter Throckmorton, would be working with me in Ceylon: see *The Treasure of the Great Reef*.

not asked any intelligent questions. Sophisticated robots might not be interested in talking to children; in any event, they would probably observe us for a few centuries before making contact.

They may, indeed, be waiting out there (inside Phobos?) until we have proved our intelligence by discovering *them*. It is an interesting exercise, and it may some day be a valuable one, to consider just how one sets about designing an electronic anthropologist. At this very moment we are building vehicles to land on the Moon and to carry out an amazing variety of automatic surveys, samplings, ground analyses and scientific measurements, but it would be an immensely more difficult task to devise a robot explorer that could cope with intelligent life. Yet it could be done.

One day, some excited astronaut from Earth may discover and board an unknown spacecraft—and promptly find himself being put through a series of tests like a rat in a psychologist's maze. Let us hope that at the end of them, the unemotional robot mind does not decide that the specimen's I.Q. is too low to be measured, but that all the same the animal is an interesting subject for dissection.

This is just one of the dangers we may encounter, if we ever tangle with alien and vastly superior technologies from beyond the stars. How long would Edison have survived, if he had been given a nuclear reactor to examine? And Edison is only two generations away from us; we may be confronted by the products of sciences thousands or millions of years in advance of our own. If you picture a boatload of Polynesian islanders finding and boarding a deserted *Queen Elizabeth,* with full steam up and all her generators running, you have a good idea of the possibilities both for discovery and disaster.

Sooner or later, I believe, we will meet whatever intelligences share our universe, and the first contact may well be through such robot ambassadors. When they have finished interrogating us they will start their engines and head out across space, to report to their unknown masters. And we will watch them go with the same doubts and misgivings that the Japanese once felt when Commodore Perry set sail; for we will know that our isolation is forever ended, and that sooner or later there will be other callers.

Fantasy? Of course; but we are moving with ever-increasing speed into an age when such fantasies will be the central facts of political life, as the race to the Moon has already become. We must stretch our minds to embrace such possibilities, and stranger ones; otherwise we can never feel at home in the incredible universe whose exploration we are now beginning.

As I watched *Oriana*'s lights dwindle away across the water like a receding galaxy, I marveled at the strangeness of being born at this unique moment of time, this fleeting moment on the frontier between two ages. For we of the twentieth century will see both the first ships of space, and the last ships of the sea.

As the aeons lengthen, and our children spread outward from Earth across the solar system, they will build the shipping traffic lanes of the future, following the planets round the Sun. Every rocket now leaving the launching pad is helping to found a tradition that may last longer than all man's dealings with the sea, and will take him to stranger ports by far than were ever visited by galleon or caravel.

And if our race survives for as long as even the least successful of the dinosaurs whom we sometimes dismiss as nature's failures, then we can be sure of this: For all but a brief moment near the dawn of history, the word "ship" will mean simply—spaceship.

The sea covers two-thirds of the planet we have mis-named Earth, and is so much a part of our lives, our traditions and our culture that we think of it as some-thing universal, eternal. Yet it is neither; it is unique to our world, and all its sagas may be no more than one brief chapter of history.

For in the beginning, there was no sea. The burning rocks of the newly formed Earth were too hot for water to condense upon them. Without a break, century after century, the greatest storms our planet has ever known raged from pole to pole, but the rain boiled skyward into steam when it touched the ground. The whole world was dry land.

And one day, the geologists and the Book of Revela-tion agree, it will be dry again. As our planet ages, it will slowly lose its envelope of air and water. The atmosphere will drift off into space; the oceans will sink down through cracks and crevices as the face of the once beautiful Earth wrinkles like that of an old, old woman.

Seas and lakes and rivers belong to the morning glory of a world, and do not long outlast its youth. We know that this is true, for there is an analogy close at hand. When we look outward from the Sun, we see a world that has already lost its oceans, for only a trace of water remains upon our neighbor Mars, locked up at the poles in a thin powdering of snow. And if—which seems unlikely—the Moon ever possessed oceans, it, too, surrendered them long ago to space.

At this moment of time, it appears, no other world knows the march of waves against the shore, the ebb and flow of tides, the white line of foam retreating down the beach. These things belong to Earth alone; the inner planets are too hot for water to exist upon them in liquid form, the outer ones, far too cold.

There is only one possible exception, and even that is not very promising. This is the planet Venus, almost a twin of the Earth in size, but some twenty-five million miles nearer to the Sun.

Venus, now the target for an increasing number of space probes and radar beams, is covered with perpetual clouds, which through the telescope appear as a blinding, featureless white. The spectroscope shows that there is water on Venus, but it may all be in the clouds, for recent radio measurements indicate that the surface temperature of the planet is far above boiling point. However, these figures are not yet final; it is still possible that Venus has seas, or at least lakes, in her polar regions or at high mountain altitudes.[1] There may even be temporary seas that condense on the dark side of the planet as it slowly rotates; we shall soon know.

Yet when we think of the word "sea," we may be taking too parochial a view. Need an ocean be made of *water?* There are other possibilities, and in the enormous, multiform complexity of the universe, many of them may be realized. On the giant worlds, Jupiter and Saturn, circling in the outer cold far beyond the orbit of Mars, there may be—indeed, we can almost say there *are*—oceans greater than any upon our own planet.

These oceans, if we can call them that, are hundreds of miles deep, and formed of liquid ammonia. They are stirred by storms so tremendous that we can see them across more than a billion miles of space. And drifting

---

[1] Even the Mariner readings, which appeared to confirm the high temperature, have been disputed by some scientists. Venus seems determined to keep her secrets to the end.

sluggishly across the southern hemisphere of Jupiter is a strange floating island as large as our entire Earth, the famous Red Spot, perhaps the only permanent feature of the planet's ever-changing surface. For Jupiter is a world without geography.

It seems unlikely that men will ever explore the turbulent, icy depths of these strange seas. But before many years have passed, our robot space probes will be descending through the atmospheres of the giant planets, braving the ammonia storms to radio information back to distant Earth.

What will they find there? Today, we do not know enough even to make intelligent guesses. Every astronomer will assure you that no form of life could possibly exist on Jupiter or Saturn, at temperatures of two or three hundred degrees below zero, and pressures of a thousand tons to the square inch. But it is worth remembering that only a century ago the biologists were equally certain that no life could exist in the depths of our own oceans.

There may even be "seas" on the Moon; if there are, they will be of dust. Some astronomers have suggested that the flat lunar plains may be covered with finely divided powder that has been flaked from the mountains by the relentless blasts of solar radiation. Dry and slippery as talcum, it could have gathered during the ages in low-lying areas, where it may be waiting to trap future explorers.

As I have suggested in a novel *A Fall of Moondust*, it could be very unpleasant stuff to negotiate. In some ways it would behave just like a normal liquid, flowing slowly under the low lunar gravity. You could walk across it with the aid of skis or snow shoes, and one day there may be paddle-wheel moon boats sailing the lunar seas, long after they are extinct on Earth. This delightfully nostalgic idea was first suggested by the science-fiction writer James Blish, and I hope that one day he receives due credit. It will also be an amusing irony if the old astronomers who gave the Latin names *Mare* and *Oceanus* to many of the dark plains on the

Moon turn out to have been not so far off the mark after all.

Much more formidable seas may exist on some parts of Mercury, the nearest planet to the Sun. It is hot enough here to melt sulphur, and possibly even such metals as lead and tin. There may be regions of the planet where the temperature never falls much below a thousand degrees Fahrenheit, for Mercury keeps one face turned always toward the Sun.

Above any valleys on the day side of Mercury, the Sun could hang almost vertically overhead forever, while its rays—ten times as powerful as on Earth—reverberated from the surrounding walls. From such valleys might flow rivers of molten metal, seeking, as do the rivers of Earth, their own infernal seas. Any mariners who ever brave these fiery oceans will require stout ships indeed.

But even molten metal is something that we can understand and can handle with techniques which date back five thousand years or more. However, at the very frontier of the solar system, almost four billion miles from the Sun, we may encounter the strangest and perhaps most terrible seas of all.

On the outermost planet Pluto, the noon temperature may occasionally soar to 350 degrees *below* zero. One the dark side, it must be much colder; the aptly named Plutonian night lasts six times as long as ours, and in the small hours of the morning it may grow cold enough to liquefy hydrogen.

We are now handling liquid hydrogen on the large scale as a rocket fuel, and it is most peculiar stuff. Quite apart from its extreme coldness—it boils if allowed to become warmer than minus 423°F.—it is extraordinarily light, having less than a tenth of the density of water. Any vessel of normal design would thus sink like a log in a sea of hydrogen; even balsa wood or cork would plummet to the bottom like lead.

Whether or not there are hydrogen lakes on Pluto is, today, anyone's guess, and it will be quite a while before we know the answer. But there is one yet stranger possibility that should be mentioned; if Pluto cannot provide it, it must surely occur somewhere in the cosmos, perhaps on a giant Jupiter-type planet that has lost its sun and been frozen for ages in the interstellar night.

The ultimate ocean—the sea to end all seas—is one of liquid helium. At the unimaginable temperature of minus 455°F.—only four degrees above the absolute zero of temperature—helium turns into a fluid called Helium II. This is a substance absolutely unique in the universe, with properties that defy common sense and even logic.

This is what could happen, if the explorers of the far future ever meet a sea of Helium II and are reckless enough to set sail upon it. To get a better picture of the improbable events that would follow, we will assume that our mariners are using an open boat.

They would get their first surprise immediately after they pushed off from the shore. Whereas on water—or any other liquid—friction brings a moving body to rest within seconds, this does not happen with Helium II. It is almost completely frictionless; by comparison, the smoothest ice is like sandpaper. The boat would therefore head out to sea with undiminished velocity, without benefit of motor. It would eventually reach the far shore—even if that were a thousand miles away—if something much more disconcerting did not happen first.

The voyagers would suddenly become aware that their boat was filling at an alarming rate. We can assume, of course, that they have already checked it carefully for leaks, and are confident that none exist. Then why is Helium II rising so rapidly above the floorboards?

Though the boat has plenty of freeboard, the stuff is coming *straight up the side and over the gunwales,* in a thin but swiftly moving film that is defying gravity. For Helium II can syphon itself from one container into another, provided that there is a connecting path between the two. In this case, the process will stop only when the level inside the boat is the same as that outside; and by then, of course, it will have sunk . . .

Let us suppose that our explorers, a little shaken, manage to get back to land and build themselves a better boat. Obviously, it must be totally enclosed, not an open dinghy, but something like a submarine. They check the hull very carefully for leaks, going over every inch of it with a magnifying glass. Not even a pinhole is visible, so they set sail again with complete confidence.

All that happens this time is that their ship takes a little longer to sink. Helium II is a "superfluid" that can race like lightning through microscopic pores and holes. Even a hull that was, for all practical purposes, airtight, would leak like a sieve in a sea of Helium II.

After these imaginary, yet possible, adventures, it is a relief to return to the familiar seas of Earth. We will never escape their call, as long as we are human. For we were born in water, and the salts of the ancient oceans still flow through our veins, though we left them half a billion years ago.

Whatever strange oceans the men of the future find upon far worlds, they will never love them as we loved the seas of Earth.

The Winds of Space

There is a wind between the worlds, whose existence was undreamed of by scientists—though not by poets—until a dozen years ago. The scientists "knew" that space was empty, containing nothing but a few stray meteors. Apart from these lonely wanderers, it was a perfect vacuum.

Well, it is not. Almost invariably, the universe turns out to be more complex than we could have imagined, and this has been true of space. By the time we have finished with it—and it has finished with us—we may well have discovered that the void between the planets is as complicated an environment as the ocean or the atmosphere. Pascal never guessed how far he was from the truth when he expressed terror of the "silence and emptiness of infinite space." Infinite it may be, but it is neither silent nor empty, except to our human senses, which have so little contact with reality.

To the radio astronomer, space is full of noises, covering octave after octave of the bands across which he tunes. Though pure noise is all that he has found so far, that has already been enough to transform our knowledge of the universe. We now know that stars, planets, and galaxies emit characteristic radio waves; we have been able to detect the hiss of these cosmic transmitters out to distances that old-fashioned optical telescopes like the 200-inch reflector on Palomar are powerless to span.

I use the word "hiss" deliberately, for that is what it sounds like to the ear: raw, undifferentiated noise, identical with that produced by any sensitive radio receiver when it is tuned between stations with the gain control full up. Only at one spot on the cosmic radio band is there a distinctive, sharply tuned signal from space. You will not be able to locate it with your ordinary radio, or even your short-wave set; it is at far too high a frequency (1,420 megacycles, or a wavelength of twenty-one centimeters) and also far too feeble. But it is there, and it is one of the most important single keys that the radio astronomers have yet found to the secrets of the universe.

This 1,420 megacycle signal is not produced by intelligent creatures on other planets. It comes from every part of the sky, and is the song of the hydrogen atoms that drift between the stars. We now believe that most of the matter in the universe—perhaps more than 90 per cent of it—is hydrogen. Much of it is concentrated into the stars, whose furnaces it fuels; but much more is dispersed among them in a gas billions upon billions of times more tenuous than the air we breathe.

In our corner of the universe, the Sun itself is the main source of this gas, which it ejects in great clouds that go scudding across the orbits of the planets. Though conclusive proof of this (as of so much else) was not obtained until the International Geophysical Year satellites started circling the Earth, one piece of evidence had pointed to the existence of such hydrogen gales, or "solar winds," for many years.

Unlike the evidence for most scientific theories, this is visible for all the world to see; it is written across the face of the sky by the tails of comets. Even those who have never seen a comet (and there have been no really spectacular ones during the twentieth century) are familiar with the appearance of these strange visitors, so well described by their ancient name of "hairy stars."

Most people, however, are probably under the impression that the tail of a comet streams behind it, like the smoke from an old-time locomotive. This is not so; indeed, the tail is more likely to point *ahead* of the comet, resembling the locomotive's searchlight rather than its smoke trail. The general rule is that, as a comet swings around the Sun—which it may sometimes do on a very tight, hairpin bend—its tail always points *away* from the Sun, whichever direction the comet itself may be moving. Thus though the tail may stream behind when the comet is making its approach to the Sun, for most of the time it is broadside on to the comet's orbit, or even pointing ahead of it.

There is only one possible explanation for this odd behavior. *Something* emanating from the Sun must be sweeping the comet's tail outward, as the wind carries away the smoke from a chimney stack. In most old astronomy books (and any astronomy book printed more than ten years ago is an old one) you will find it stated with considerable assurance that the pressure of sunlight is the agent responsible.

It is rather hard to realize that sunlight can exert pressure, for what could be more insubstantial than light? But as everyone is well aware since Hiroshima, energy and mass are two sides of the same coin. Light, therefore, has mass and hence momentum; if you hold out your palms toward the sun on a bright clear day the weight of light falling upon them will be about a millionth of an ounce.

It seemed reasonable to assume that out in space this force, weak though it is, might be enough to affect the extremely thin gas composing a comet's tail. However, calculations show that though sunlight may have some minor influence, its pressure is much too feeble to produce the spectacular changes often observed in comets. Occasionally the tail of a comet is torn off by a blast of invisible energy which must be hundreds of times more powerful than mere sunlight. We have, therefore, to look for another explanation, and we find it in the fact that the Sun emits other things besides light. It is the solar wind which carves and shapes the tails of comets.

This wind is electrified, as is all the matter in so intensely hot an object as the Sun. The hydrogen atoms have been torn asunder, and their two components—the central positive proton and its orbiting negative electron —set moving so violently that they cannot recombine. They are like partners who have broken up a sedate waltz and gone spinning off independently in a wild bacchanal.

Such a gas, because it contains equal numbers of positive and negative charges, is itself electrically neutral: It is known as a plasma—a term you will be hearing more and more often in the future, for it represents a state of matter which is becoming of increasing technical importance. You have a sample in your own home, if you employ fluorescent light tubes.

This plasma wind varies with the activity of the Sun in a manner which as yet we cannot begin to predict, or even to explain. We know, however, that it originates from the most violent of all the spectacular events that take place on the Sun's surface—solar flares.

A flare is a sudden local brightening of the Sun's face—usually in the neighborhood of a sunspot—which may within minutes spread over hundreds of millions of square miles. It lasts from thirty minutes to a few hours; during this short period of time, an area equal to that of a hundred earths may blaze forth with such brilliance that it outshines even the surrounding incandescence. The energy released by a flare is beyond all imagination, sometimes equaling that of a million H-bombs, and it is not surprising that such explosions blast planet-sized clouds of the Sun's atmosphere into space.

These clouds reach the Earth about a day after they are ejected from the Sun, and their arrival triggers a very complex train of events. Most of that complexity is due to the fact that the Earth possesses a magnetic field which acts like a giant trap, thousands of miles across. Just as iron filings are attracted to the poles of a magnet, so the plasma clouds from the Sun are funneled down toward the magnetic poles of Earth. When, as often happens, they eventually enter the atmosphere, they make it glow with the wonderful auroral hues that have amazed and baffled men for centuries. Thus even in the depths of the polar night, the colors flickering over the landscape still have their origin in the hidden sun.

At this point, it may be as well to give some more of the evidence in support of these statements. Much of it is highly technical, but two recent experiments have given direct and dramatic proof that plasma winds do in fact blow from Sun to Earth.

The first involved Pioneer V, the deep-space probe launched on March 11, 1960 into an orbit toward Venus. Pioneer V was tracked out to a distance of twenty-two million miles, and was the first man-made object ever to send back information from the deeps of interplanetary space. It is still orbiting, its radio voice now silent, between Earth and Venus; perhaps in the far future it may be recovered and displayed in some museum of astronautics as a quaint relic from the past.

Twenty days after launching, when it was three million miles sunward of Earth, Pioneer V ran into a great cloud of electrified gas. A little later, that cloud reached Earth, where it was noted by satellites just outside the atmosphere, as well as by instruments at ground level. In a few years, there will undoubtedly be space probes on permanent patrol inside Earth's orbit, to give warning of plasma clouds heading in our direction; as we will see later, such information will one day be literally a matter of life and death. Even now it would be of great value to the radio and cable companies, for these clouds can cause communication blackouts all over the world.

The other evidence comes from Echo I, the most brilliant of all the satellites, which millions of people saw as a brightly moving star. Echo I, a hundred feet in diameter, is merely an inflated balloon and so has extremely small mass for its size. Like a drifting soap bubble, it responds to the slightest influences. In fact, it is so flimsy a structure that even the feeble pressure of sunlight has had a very substantial effect on its orbit, changing its distance from Earth by hundreds of miles.

This light-pressure effect is a steady push that builds up uniformly day after day. Occasionally, however, Echo I has been buffeted by sudden disturbances which produce sharp changes in its orbit—changes far greater than could be caused by the gentle push of light. These are due to the impact of solar plasmas; though earlier satellites showed similar effects, Echo I is the first to be blown badly off course by the winds of space. It will not be the last.

All this may have given you a false idea of the actual strength of the solar wind, so let us compare it with the phenomena of everyday life. It is rather difficult to do so, because the forces and quantities concerned are so far outside the range of ordinary experience that the figures tend to be meaningless: for example, even in the most violent hurricane, wind velocities seldom exceed a hundred miles an hour.

The winds of space blow a little faster, since they can cross the 93 million miles between Sun and Earth in about a day. This gives an average speed of 4 million miles an hour, but they can probably exceed this on occasion.

Obviously, if the solar wind had a density even remotely approaching that of terrestrial air, it would not merely blow an Echo satellite off course; it would blow Earth out of its orbit, and the whole solar system would rapidly come apart. Luckily, even the densest plasma cloud from the Sun is so incredibly tenuous that it contains only about two million hydrogen atoms in a volume the size of a matchbox.

Perhaps you think that this is a respectable number, but in a matchboxful of ordinary air, there are not two million but two *thousand million million* million atoms. Or, to put it in a way which is perhaps a little more meaningful, you would have to let a matchbox full of air expand until it was a cube two miles on a side before its density had dropped to that of the solar plasma. And this is a peak value, met only when the Sun is active; the normal density is a hundred times smaller than this. It is not surprising, therefore, that the forces produced are extremely small; when you are walking into a gentle breeze, you are fighting a million times the pressure that a man in a space suit would experience if he tried to buck the fiercest wind that blows from the Sun.

Yet feeble though they are, those winds can be deadly —and now we come to what may well be their greatest importance in the future. They are responsible for at least one of the huge radiation belts that surround the Earth, which would-be astronauts are now regarding with some alarm.

When it is trapped in the Earth's magnetic field, the solar plasma forms a gigantic, doughnut-shaped cloud, the outer Van Allen belt. (The smaller, inner belt has a different origin, which does not concern us here.) The discovery of this radiation zone was one of the first, as well as one of the most remarkable, results of the International Geophysical Year. It is hard to realize that to a being who could see by radio waves, our planet would appear surrounded by structures more complex than the rings of Saturn. Yet to our eyes, they are completely invisible, and until the first American satellites started to probe them, scientists had no knowledge of their existence.

These vast smoke rings of electrified gas girdling the world high above the equator are little danger to outward-bound space travelers, who will flash through them in minutes, but it will be a different matter for man-carrying satellites. The weight of shielding required to protect the crews may make manned space stations impractical in the denser layers of the Van Allen belts. We may have to employ robots here, putting our human observers at higher or lower levels, in the gaps between the radiation zones.

There is, however, one other possibility. If the Van Allen belts prove to be a nuisance, *we may remove them.* This startling (indeed at first sight almost megalomaniac) idea was first put forward by the physicist Professor Fred Singer, who pointed out that as the total amount of material in the belts is extremely small, it would be theoretically possible to neutralize it. This might be done by suitably designed satellites which, after a few hundred or thousand orbits, would sweep up the unwanted particles and make the immediate vicinity of the Earth more fit for human occupation. After air conditioning, space conditioning . . .

That this idea is no fantasy was demonstrated in the summer of 1958, when the Advanced Research Projects Agency of the U.S. Department of Defense conducted what has been, with some justice, called the biggest scientific experiment in history. This experiment—Project Argus—involved not the neutralization of the Earth's radiation belts, but something still more remarkable: *their creation.*

The philosophy behind Project Argus was as follows.

If the Sun could produce radiation belts around the Earth by blowing charged particles into the upper atmosphere from space, we should be able to do the same thing by injecting them from beneath. Calculations revealed the surprising fact that even a very small atomic bomb could provide enough plasma to give observable effects.

This suggestion, incidentally, was first made by a rather remarkable man, a Greek electrical engineer named Nicholas Christofilos, who some years ago caused considerable embarrassment to United States physicists by suggesting a way in which several million dollars might be saved in the building of their giant accelerators or atom smashers. Because it was couched in somewhat unorthodox mathematics, his letter went into the "Nuts" file. Here it remained until the principle put forward by Christofilos was independently discovered and announced, and a modest cough from the author drew attention to his priority. This is the sort of incident that can lead to recriminations and lawsuits, but in this case there was a happy ending: Christofilos, despite his lack of the right academic qualifications, was imported into the United States and is now at the University of California's famous Radiation Laboratory.

Project Argus was carried out in great secrecy, as it had profound defense and political implications. On three days in August and September 1958, very small nuclear devices were launched to a height of three hundred miles over the South Atlantic by a U.S. task force, and injected huge numbers of electrons into the upper atmosphere.

Within seconds, the results were noted by ground radar stations, satellites and instrument-carrying rockets in many parts of the globe. Still more impressive is the fact that they were also observed by the naked eye. The August explosions produced spectacular auroral displays, both in the detonation area and thousands of miles away, where the Earth's magnetic field focussed the electrons back into the atmosphere. *For the first time, men had made an artificial aurora.*

The plasma from the explosions lasted for many days; indeed, satellite borne instruments were able to detect some of the Argus electrons three months after they had been shot into orbit. The experiment showed, beyond all doubt, that we now have the power to produce really striking changes in the character of nearby space. Only the future will show what use we make of these powers; a couple of years before Project Argus, in a short story ominously entitled "Watch This Space," I drew attention to the ghastly possibilities of cosmic advertising, and I only hope that Madison Avenue does not lure Mr. Christofilos away from the Radiation Lab.

The Argus experiment was important because of the information it gave about the natural radiation in space; once you can reproduce a phenomenon, you are well on the way to understanding it. Such understanding will be vital when men start to travel between the planets, for not only Earth, but possibly all the major bodies in the solar system, may have radiation belts. Indeed, there is already some evidence that Jupiter possesses a radiation zone thousands of times more intense than Earth's. We will have to chart these regions and, if necessary, minimize their danger by racing through them at high speed.

The planetary radiation belts probably wax and wane with solar activity, which rises to a peak every eleven years. The Sun was at its maximum in 1958, when the Van Allen Zone was discovered, and flares were occurring on its surface at least once a month. Now it is sinking into quiescence, but it will revive again around 1969.

And this is an alarming thought, for just about then manned space flight will be really getting under way. When men start to sail away from the Earth, they will be doing so under the worst possible conditions—into the teeth of the storms of space. The indications are that no practical weight of shielding could protect the crews of a spaceship from a really severe blast of solar plasma. They would feel nothing and see nothing as that impalpable wind swept past their vessel, yet in a few hours they would start to die of radiation poisoning.

What is the answer? It is much too early to be sure. If worst comes to the worst, we will have to travel only when the Sun is quiet. When it is spewing forth its deadly clouds, our ships must remain behind the shield of the Earth's atmosphere, protected as if by a breakwater from the gales of space. Since a careful watch on the Sun for solar flares will always give us a day's warning of an approaching storm, this will be no great handicap for lunar travel, for flights to the Moon will last no longer than this.

For ships in deep space, weeks or months away from Earth, it will be a different matter. They could not run for shelter—there might be none within ten million miles—but would have to ride out the storm. Though it would be impossible to provide shielding for an entire ship, the individual crew members might be supplied with thick-walled capsules or storm cellars within which they could remain, cramped but safe, until the danger was past. Eventually, there can be little doubt, we will find ways of deflecting solar plasmas by magnetic or electric fields. The energy required to do so would be very small, though its application is beyond our present-day technologies.

There are many analogies between the seas of Earth and the vast seas of space; here is a final one. Two centuries ago, when men set out on long ocean voyages, they were attacked by an obscure and horrible disease which often wiped out entire ships' companies. Perhaps radiation sickness will be the scurvy of the first astronauts; its cure, however, will demand more than a few limes and lemons. But we will find it, sooner or later, so that we can safely ride the winds of space.

How strange it is to think that, since the beginning of time, these invisible winds have been sweeping past our world, gusting from Earth to Moon in fifteen minutes, yet until a moment ago, no man dreamed of their existence. What other surprises are waiting for us, out there in what we once believed was empty space?

Nothing in our lives is more basic, more unavoidable, than the ceaseless rhythm of night and day, or the regular progress of the seasons as the Earth makes its annual journey round the Sun. From the beginning of history, men have been conscious of time, and haunted by its meaning. It is not too fanciful to say that we have been born and bred inside a cosmic clock whose face is the pattern of the stars, and whose hands are the Sun and the Moon. We have never been able to escape from it—until now.

It is about time (no pun intended) that we look at some of the consequences of this escape from our particular cog in the celestial clockwork, the planet Earth. For we will soon have to learn that there is nothing fixed or sacred about days and weeks and years; not only do they differ from world to world, but there are even planets upon which such things will have no meaning.

Before we consider the strange clocks and calendars that the travelers and tourists of the future will need, let us have a brief glance at timekeeping here on Earth. We take date and time so much for granted that we often forget that it required thousands of years, and some of the finest minds the human race has ever produced, to establish our present system of timekeeping or chronology.

The basic unit of time is, of course, the day. The concept of the day is not merely obvious, it is unavoidable.

Even the simplest animals are aware of the difference between light and darkness, though it took mankind a great many centuries to discover that the effect was due to the spin of the Earth on its axis, *not* to an actual movement of the Sun across the sky.

Quite arbitrarily, we divide the day into twenty-four hours. Let us be absolutely clear on this point. Hours (and their subdivisions, minutes and seconds) are the inventions of man, and have no counterparts in the natural world. We might say that God made the day, and the rest is the work of man. The day might just as well have been split into ten or a hundred parts, though twenty-four has certain purely numerical advantages, since it can be conveniently divided into thirds, quarters, sixths, eighths and twelfths.

Equally arbitrary are the week and the month; were it not for the accident that we have a single bright Moon that takes a little more than seven days to change from quarter to quarter, and nearly thirty to go through the full cycle of its phases, we might not have settled upon either interval. As the Earth is the *only* planet with a single moon (all the others have none or several) it is quite clear that the month is quaint local custom; but it is a convenient one which we will probably export elsewhere.

The year, on the other hand, is quite another matter. It is as fundamental, as inescapable, as the day, and dominates human, animal and plant life to an equally marked extent—especially in high latitudes, where summer and winter bring such changes that they might almost be in different worlds. Even near the equator, where there is little change of temperature with the seasons, the cycle of the year still makes itself felt, for it brings such regular variations of wind and rain as the famous "monsoons." Nowhere on Earth can we escape from the effects of our planet's annual march around the Sun.

The day and the year are, therefore, the two natural and basic intervals of time; we have to accept them as they are and somehow incorporate them into our sys-

tems of chronology. All would be well—or fairly well—
if the day went into the year an exact number of times.
Unfortunately it doesn't, and this stubborn fact has
given calendar makers headaches for at least five thousand years.

Probably the first rough estimate, soon after men
learned to count at all, was that the year lasted 360
days, which gives a beautifully simple calendar of twelve
30-day months. Unfortunately, such a calendar goes
haywire very rapidly, because the Earth actually takes
a little more than 365 days to complete one circuit
around the Sun. If you start counting off 360-day
"years," therefore, you get out of step with the seasons
at the rate of five days every year. After only thirty-six
years you're six months out of phase and are celebrat-
ing Christmas at midsummer—unless you've starved to
death first because your crop planting has been equally
in error. For an accurate calendar is not merely a social
convenience; it is a matter of life and death. Agricultural man knew this very well, as he watched the star
patterns that marked the coming of spring; but super-
market man tends to forget it.

A 365-day calendar is a great improvement, and a
365¼ day one is better still. This, of course, is the one
we use, by the simple dodge of inserting an extra day
every fourth (leap) year. If the Earth took precisely
365 days and 6 hours to go around the Sun, this calendar would be correct forever. As the year is actually
a little shorter than 365¼ days, additional adjustments
are necessary from time to time, but only at very long
intervals. (Still, it's amazing how these errors can add
up, like dust swept under the carpet, if you don't do
something about them. By 1752 the calendar was no
less than 11 days out of kilter, and it was therefore necessary to decree that September 2 would be followed
by September 14. I suspect that most politicians would
be very embarrassed, even now, if they had to face
the cry that then went up from the indignant populace:
"Give us back our eleven days!")

So much, then, for the basic principles of time measurement on the planet Earth; now we are ready to face the peculiar things that happen off it. Peculiar to *us,* that is; to other beings, if they exist, we will be the oddities, with our comic clocks and calendars.

The first astronauts have already encountered some of the problems of timekeeping in space, for they have experienced a ninety-minute day. This is the time that a close satellite, like a Mercury capsule, takes to go around the Earth. Every hour and a half, therefore, a traveler in such a capsule witnesses a sunrise; every hour and a half he sees a sunset. He learns, as no men have ever done before, that daytime and nighttime are private experiences; they depend on where you are, and how fast you are moving. Night itself is purely a local effect—the temporary eclipse of the Sun by the shadow of a planet. The further our orbiting astronauts travel into space, the less they will see of night. When they are thousands of miles out, they will hardly ever pass through Earth's narrow cone of shadow; they will be in eternal light—perpetual day. They will always see the Sun shining in their sky.

Space travelers en route will be in much the same situation as submarines during a prolonged undersea voyage. They can use whatever method of timekeeping suits their convenience; whether it is day or night outside the walls of their vessel scarcely concerns them. Almost certainly, astronauts will adopt Greenwich Mean Time, as thousands of navigators at all heights and depths all over the world are doing at this very moment. Spaceships will carry incredibly accurate clocks, measuring the passage of time by the unvarying vibrations of atoms, not the oscillations of balance-wheels, which are subject to all sorts of irregularities. Such atomic clocks can measure time to an accuracy of a few parts in a million million; it would take tens of thousands of years for them to gain or lose a single second!

The simple solution of using GMT, unfortunately, breaks down as soon as we land on any other planet. For then we will no longer be living in our own private little world, and it becomes essential to have clocks and calendars which bear *some* relation to the natural events taking place around us. The scientists may stick to GMT, but future planetary colonists would like to have clocks that will at least tell them whether the Sun is shining out of doors. And this is where the trouble begins.

It will start first on the Moon, though the situation there is not as bad as it might be. Since Earth and Moon form a single closely knit little family, they go around the Sun together and so have the same year. The lunar "day," however, is extremely long—just over twenty-nine and a half of our days.

It is hardly likely that human beings will ever adapt themselves to fifteen days of wakefulness and fifteen days of sleep, nor will it be necessary. But it will be necessary to have clocks which show the slow progress of the sun across the lunar sky, and you may be surprised to know that such time pieces can be bought at any good watchmakers. You may even be wearing one at this very moment.

Those watches that show the phases of the Moon will be very handy *on* the Moon. They could be easily set so that the changing crescent gave an approximate answer to the vital question, "How many days till sunset"? And on this side of the Moon there would, of course, be another timekeeper always visible in the sky. As it went through its phases—the exact reverse of those that the Moon shows to us—the huge and brilliant Earth would tell the experienced traveler just what time of lunar day it was. For when the Earth is full, the Moon is new, and vice versa.

It is when we get to Mars that things start to become really complicated. In the first place, the Martian day is not the same length as ours, though it is surprisingly close to it—about twenty-four hours and forty minutes. This is not too bad as far as working schedules are concerned, and it would be simple to modify an ordinary watch so that it ran slow enough to keep good Martian time. (Before you try it, I should mention that the usual regulators seldom allow you to lose the necessary forty minutes a day. Few watchmakers expect their products to be *that* much in error.) The explorers would soon adapt themselves to the slightly stretched twenty-four hour clock, and even that sensitive time-keeper, the human stomach, would not be inconvenienced by the extra five or ten minutes between meals.

We cannot, however, dispose of the Martian calendar so easily. It takes almost two Earth years (to be more precise, 686.979702 . . . days) for Mars to make one revolution around the Sun. But we're not concerned with the number of *Earth* days in the Martian calendar —only the number of *Mars* days, which is obviously a little smaller. It's actually 668.599051. What, you may well ask at this point, will a 668.599051-day calendar look like?

Well, at least two astronomers—Dr. R. S. Richardson of Mount Palomar, and Dr. I. M. Levitt of the Fels Planetarium, Philadelphia—have attempted to construct one. Assuming that we stick to twelve months so that we can retain the familiar names January, February and so on, each quarter of the Martian year might contain one 55-day and two 56-day months. This is much neater than the messy calendar we have inherited from Julius Caesar and Pope Gregory XIII, though it would be rather a long time between paychecks.

Four quarters of 167 days each give a 668-day year, which is almost exactly 0.6 days short of the actual Martian year. To correct this, we need rather a lot of leap years—obviously, six in every ten, or three in every five, as compared to one in four on Earth. However, this calendar would be accurate over extremely long intervals; no further corrections would be needed for a thousand years.

Reconciling terrestrial and Martian calendars will be a complicated and tedious job, and no doubt the interplanetary diaries of the future will contain tables so that you can compare dates at a glance. Dr. Levitt has done better than this; in cooperation with the Hamilton Watch Company, he has designed a clock to show both the time and the date on Earth and Mars. When this clock was built several years ago, I am sure that neither the doctor nor the company realized how soon it would be needed.

Incidentally, from what point in time will we start reckoning the Martian calendar? It will make the task of future historians easier if we begin Year I with the first landing. This would also reduce the risks of confusion between the two calendars. A date like March 14, 5 will hardly belong to Earth—and there would be no doubt at all in a case like February 47.

But even with the greatest care on both worlds, there are bound to be difficulties and mistakes as Earth-Mars traffic and trade increase. Here are some of the problems that lurk in ambush for the next generation, and perhaps for some who read these words, since the first landing on Mars cannot be much more than twenty years away.

By the Mars calendar, fifty will be a ripe old age, since it will correspond to ninety-four years on Earth. Martian colonists will be able to vote at eleven, marry at nine. A change of residence from Earth to Mars, or vice versa, would produce frightful complications. Although the biological rate of aging will, we assume, be the same on both worlds, a colonist returning to Earth would start clocking up birthdays almost twice as quickly as if he had stayed on Mars. Tourists of the future will simply have to get used to the fact that calendars, as well as currencies, can have different rates of exchange. The lawyers are going to have a field day drafting time clauses in interplanetary contracts.

Even when they stay at home and don't try to do business with Earth, the Martians (let's drop that snooty "colonial" tag, shall we?) will have troubles of their own. The long months and the numerous leap years have already been mentioned, but one could grow accustomed to these. A subtler problem is one which was first encountered by terrestrial explorers more than four hundred years ago, and is probably destined to cause many headaches on Mars.

On September 7, 1522, the exhausted survivors of Magellan's expedition landed in Spain after circumnavigating the globe for the first time in the history of mankind. As soon as they did so, they made a horrible discovery, doubly distressing to good Catholics who paid due regard to Saints' Days. By their careful reckoning, it was only the 6th, not the 7th, of September; *somewhere*, they had lost a day. . . .

I am by no means sure that, even now (and even after seeing *Around the World in Eighty Days*, where Phineas Fogg *gained* a day by traveling in the opposite direction from Magellan) many people could give a clear explanation of what happens when one crosses the International Date Line. (Don't worry; I'm not going to give one, either.) We all know that the thing exists, and that it has been safely parked in the middle of the Pacific Ocean so that it causes the minimum of inconvenience to everyone—except to travelers jetting between Asia and America, who frequently find they've arrived before they started.

The Martians—some of them, anyway—will be in an even worse plight. As there are no oceans on the planet, *their* International Date Line will have to lie on land. This means that, somewhere on Mars, sometime in the future, there will be a street or a square or a city park where it's Friday on one side and Saturday on the other. It will be quite a tourist attraction, but a great nuisance to the people who have to live with it. Can you imagine what it would be like in your town, if the date changed when you crossed the road? The fact that this doesn't happen in some residential area on Earth is pure luck; it must occur on *all* inhabited planets that do not possess convenient oceans in which unwanted days can be drowned. I have used this idea in a short story, "Trouble with Time," which you will find in *Tales of Ten Worlds*.

Yet if you think that the Martians have problems, wait until you get to Jupiter.

So far, the worlds we've considered all have one thing in common. Their "days" may be of different lengths, but at least they are the same everywhere on the planet. This is no longer true on Jupiter, for this giant among the Sun's children does not revolve as a solid body—it turns more rapidly at the equator than at higher latitudes. If you lived in the Jovian tropics (not that they are very tropical, almost half a billion miles from the Sun) you would find that the day lasted about nine hours fifty minutes of our time. But as you went north or south toward the poles, it would lengthen by approximately five minutes.

This means, of course, that Jupiter has no fixed geography; its surface is plastic so that regions in higher latitudes slowly drift backward as the planet spins. To imagine such a state of affairs on the Earth, you must picture a situation in which Canada and South America steadily move westward with respect to the United States. This may indeed actually happen, but it takes geological ages—not millions, but *hundreds* of millions, of years—for the effect to be noticeable. On Jupiter, it requires only a few months, so both timekeeping and map making on the planet involve fearful complications.

Fortunately, these problems may never be of much practical importance, for Jupiter is one planet that human explorers will probably be glad to leave alone. An atmosphere of hydrogen, methane and ammonia, at pressures rising to millions of tons to the square inch and ripped by thousand-mile-an-hour winds, is not an attractive prospect, especially when coupled with a gravity two and a half times Earth's and temperatures approaching three hundred below zero. Any tourist agency preparing a travel folder on Jupiter will have to overwork that useful adjective "bracing."

However, even if we leave the exploration of Jupiter to robot space probes, we will certainly land on the planet's dozen or more moons, several of them large enough to count as worlds in their own right. (Ganymede and Callisto, for example, are almost as big as Mars.) As they all accompany Jupiter on his journey around the Sun, they will share his year, which is eleven years ten months of *our* time. A single "month" of Jovian time would thus last as long as an Earth year.

As we move still further outward beyond Jupiter, the planetary years become ever longer, approximately doubling as we jump from one world to the next. Saturn takes 29½ years to complete its circuit around the Sun; Uranus, 84; Neptune, 165; and Pluto, the most distant of the planets, no less than 248.

On such remote worlds, the conception of the year can have little meaning. There will be no seasons, for the temperature is always hundreds of degrees below zero. The Sun is little more than an exceptionally brilliant star, providing plenty of light, but no heat. On Pluto, the noonday sun would be about as warm as our full moon.

The "days" of the outer planets present us with a curious paradox. The larger the world, the shorter its day—not, as you might reasonably expect, the other way around. We have already mentioned Jupiter's flexible 9 hours 50 to 9 hours 55 minute day; Saturn revolves in 10 hours 14 minutes, Uranus in 10 hours 49 minutes, Neptune in (probably) 14 hours. Little Pluto, on the other hand, has a slow day, six and one half times longer than ours.

I suspect that for the giant planets and their many satellites, we will stick to earthly calendars and clocks, just as we will do aboard our spaceships. Any settlements we may establish on the outskirts of the solar system must, of course, be completely sealed from their ferociously hostile surroundings. The rapid daily sweep of the Sun across the sky, and its slow annual crawl around the zodiac, will have none of the significance that they do on Earth.

It will be far otherwise when we travel in toward the heart of the solar system. On the two inner planets, Venus and Mercury, the Sun is the dominant factor in life and its movements must be all-important. As on Earth, Moon and Mars, we will once again have to adjust our clocks and calendars to local conditions.

Our next-door neighbor Venus has a short year of

225 days. The Venusian (Cytherean? Venerian? Take your choice) day is, however, a major mystery. Because the planet is perpetually veiled in clouds, no definite surface features have ever been observed from which the rate of rotation can be deduced. Indirect evidence suggests that Venus takes at least ten Earth days, and perhaps considerably longer, to rotate on her axis. If this is the case, there will be less than twenty Venus days in the Venus year. There may be only one. However, the observations of Mariner I indicate that the atmosphere of Venus is so dense and opaque that no sunlight ever reaches the almost red hot surface of the planet. In that case, we have a situation like that in the depths of our own oceans. More than a mile down in the sea, by day or night, there is perpetual darkness. The abyss knows nothing of the Sun; so it may be with Venus.

On Mercury, the planet nearest to the Sun, we will encounter a situation that is almost baffling in its simplicity. This welcome change arises from the fact that the planet keeps one side always turned toward the Sun, so that the very conceptions of night and day, sunrise and sunset, have no meaning there. On one hemisphere there is eternal light, on the other, eternal darkness. The only measure of time that has any significance on Mercury is the year, which lasts a mere eighty-eight Earth days.

It is hard for us to imagine a world which will know nightfall only when the Sun itself expires. If you wish to see night on Mercury, it is no good *waiting* for it—you must travel to the dark side of the planet, until the Sun sinks below the horizon.

And there you will meet a night indeed, unmatched elsewhere in the solar system. The Mercurian night has lasted, in all probability, since life first emerged upon the face of our planet. It must be cold beyond imagination, not far above the absolute zero of temperature. Yet it will be a night glorious with stars and dominated by two beautiful apparitions—the electric blaze of

Venus and the scarcely less brilliant double-star formed by the twin worlds of Earth and Moon.[1]

We have now completed the roll call of the solar system, apart from such unimportant bodies as comets and asteroids. But we have scarcely begun to examine all the possibilities that exist in this incredible universe of ours, so I would like to look at one more astronomical situation—perhaps the hardest of all for our minds to grasp.

Whatever their individual differences, the planets of *this* solar system do have one thing in common: they all go around the Sun, and so each experiences a year of a fixed and definite length ranging, as we have seen, from 88 days in the case of Mercury to 248 years in the case of Pluto.

The key phrase in that last sentence is: *they all go around the Sun*. But the planets of other solar systems may not go around a Sun; they may have half a dozen from which to choose . . .

For our Sun is a cosmic recluse, wending its lonely way millions of millions of miles from its nearest stellar neighbor. Most stars, on the other hand, are sociable, occurring in pairs, triplets or even more complicated family groups. Any planets of such stars cannot move in simple, approximately circular orbits that repeat themselves regularly, age after age. The conflicting tugs and pulls of the ever-changing gravitational fields in which they move make this quite impossible.

The planets of multiple star systems must travel on strange, looping curves of inconceivable complexity that will never repeat themselves again as long as the universe endures. Look at the curves around the border of a dollar bill; the mathematics of those is utterly childish in comparison.

When there are many suns in the sky, crawling along different paths and rising and setting at different times,

---

[1] Very recent radar observations, still to be confirmed, suggest the astounding facts that (1) Venus's *day* is longer than its *year;* and (2) Mercury may after all be slowly rotating with respect to the Sun, so that its day and year are *not* identical.

it is no longer possible to speak of either days and years; the terms are as empty of meaning as color to a blind man. The inhabitants of such worlds, should any exist, must invent mechanical devices if they wish to measure time. There will be no celestial clocks in the heavens, no regular and predictable movement of a single sun and a single moon across the background of the stars. The very concept of time might never be discovered on such worlds. And perhaps, because of that, their inhabitants will know nothing of our obsession with change and decay; they will neither dream of the future nor yearn for the past. Living in an eternal present, they will have conquered time—by ignoring it.

The Playing Fields of Space

Space travel will not be all work and no play; wherever men go they must have relaxation, physical or mental. Though it may seem a little premature, not to say frivolous, to spend much time discussing space-age sport, the subject turns out to be highly instructive. It is also full of surprises, for the new gravitational and physical conditions beyond the Earth will not only transform many existing sports, but make possible fantastic new ones.

By an irony that the crews will hardly appreciate, there will be very little space in the first spaceships; it will all be outside the walls, and everyone will be most anxious to keep it there. During their off-duty hours pioneer astronauts will have to relax with cards and chess (checkers, we must assume, will be beneath the dignity of men who will probably average two doctor's degrees apiece). Not until fairly large bases are established on the Moon and planets, and in satellite orbits, will space sports really come into their own.

The Moon, thanks to movies and comic strips, already has a certain cozy familiarity. As it is an utterly airless world, the first explorers will have to wear space suits when they leave the shelter of their ships. These suits will be elaborate and bulky affairs, for they must not only supply the wearer with oxygen but must also provide protection from the fierce temperature extremes that exist on the Moon—where a single step from sunlight into shadow may bring the thermometer tumbling four hundred degrees. No one wearing a space suit will feel very athletic; but sooner or later very large areas of lunar landscape will be enclosed—either by rigid domes or flexible, air-supported structures—and provided with artificial atmospheres. When this happens, the colonists (they will no longer be pioneers) will be able to discard their clumsy pressurized suits and will be able to move around unhampered.

They will do so with a dreamlike ease that we on Earth may well envy, but will never be able to emulate. For the gravitational tug of the Moon is only one sixth as powerful as that of our planet; a man who weighs 180 pounds here will weigh just under thirty pounds on the Moon.

All objects thrown, tossed, shot or otherwise projected on the Moon will travel six times as far, and rise six times as high as they would on Earth. A high jump on the Moon would thus be a spectacular performance, though not quite as spectacular as you might think. The present terrestrial record is just over 7 feet, but this does not mean that a lunar high jumper could do six times this, or 42 feet. When an athlete clears a 7-foot bar, he actually hoists himself less than 5 feet; his center of gravity, which is around waist level, is already some 3 feet from the ground. Allowing for this, the high-jump record for the Moon will be around 30 feet, and the whole performance will take almost ten seconds. The broad-jump record, now about 27 feet, would become more than 150 feet on the Moon.

The various objects which athletes like to hurl will cover correspondingly greater distances on the Moon. There won't be enough space, unfortunately, in our pressurized lunar cities, for throwing the discus (the terrestrial record would correspond to a lunar 1,175 feet), the javelin (1,500 feet), or the hammer (1,280 feet). Even putting the sixteen-pound shot—which will become a two and a half-pound shot on the Moon—will strain the available accommodation. It will travel close to 375 feet.

If anyone ever drives a golf ball on the Moon, he'll have to put a radio transmitter inside it to track its flight. (A perfectly practical idea, by the way.) Even on Earth, golf balls have traveled 1,290 feet. A good drive on the Moon could cover at least 10,000 feet. Not only the low gravity, but the absence of air resistance, would help to give the ball an enormously increased range. All that this proves, of course, is that if we do transfer conventional sports to the Moon, we'll have to change the equipment. I dare not imagine what a St. Andrews pro would say about golf balls stuffed with lead, but we may be forced to some such solution.

There are certain games, however, that would be virtually unaffected by change of gravity. These are games that depend not upon weight, but upon mass or inertia. The two characteristics are very frequently confused, though they are really quite distinct.

The weight of a body depends entirely on the gravity field it happens to be occupying. That's why the same object can weigh a pound on the Earth, a sixth of a pound on the Moon, twenty-eight pounds on the Sun and nothing at all in an orbiting satellite. But its mass —by which we mean the opposition it gives us when we try to set it moving—is absolutely independent of gravity, and is the same throughout the universe. An object that would be weightless in orbit requires the same effort there as on Earth to set it in motion.

We can make a list, therefore, of the games that can

be played on the Moon and planets without any alteration to the rules or equipment. They're all games that involve rolling, sliding or bumping, but not throwing or projecting. Two obvious examples are bowling and billiards; and I'm sure that Lewis Carroll, who described the most famous croquet match in all fiction, would have loved the idea of transferring this gentle game to the Moon. It would work fine there.

It hardly seems worthwhile going to the Moon to play croquet; however, there is one lunar sport that may some day become a major tourist attraction. On the Moon, inside the air-filled domes that the future colonists will erect, a man could fly like a bird. It would be relatively easy, and would probably require little more than batlike wings attached to wrists and ankles. With these, we could enjoy during waking hours an experience we have known so far only in dreams.

Muscle-powered flight opens up a whole spectrum of sports and games, from straightforward racing to an aerial equivalent of water polo. I can see the time coming, not more than thirty years from today, when the TV channels will be dominated by sportscasts from the Moon. The sluggish and leaden-footed sports of Earth will seem tame compared with those that could be played on the Moon.

Everything I have said about the Moon applies, in a slightly less exaggerated degree, to Mars. This is the only planet upon whose surface we may be able to venture without elaborate protection. Mars has a thin atmosphere (though not a breathable one) and during the daytime it is sometimes comfortably warm. We may be able to manage there with simple oxygen masks, though this is by no means certain. All the other planets are much too hot, or much too cold. (Even Venus, which looked promising a few years ago, now turns out to have a temperature of about 600 degrees. This is rather a blow as I had hoped to go skin diving there.)

The gravity of Mars is about a third of the Earth's, or twice the Moon's. The lunar records I have quoted above can thus be divided by two to indicate what we may expect to achieve on Mars. It's a pity we cannot practice there first, and work our way up to lunar standards. I am afraid a lot of people will break their necks on the Moon, attempting those high jumps and coming down headfirst.

The Moon and Mars, and the major satellites of the giant planets, are fairly large bodies with respectable gravities. But besides these there are thousands of pint-sized moons and asteroids (minor planets) in the solar system. Some very peculiar things could happen on these.

Consider little Phobos, the inner moon of Mars. It is a chunk of rock about ten miles in diameter—at least we think it is rock, though a Russian scientist has recently given plausible reasons for believing that Phobos may be an artificial satellite put up by the Martians a few million years ago. In any event, its gravity pull is tiny; a simple calculation shows that a man standing on Phobos would weigh about four ounces.

This is practically, but not quite, the same as no weight at all. A stone would take a minute to fall sixty feet, instead of the two seconds it requires on Earth. Such a state of affairs is almost impossible for us to imagine, yet this is what would happen on a world like Phobos. Anyone attempting a really high jump would be in grave danger of never coming down. Indeed, it would be possible to jump clear off the world—to reach the "velocity of escape" by unaided muscle power. Our own planet's velocity of escape, so far achieved only by a few spaces probes, is 25,000 miles an hour. But on minuscule worlds like Phobos, speeds under twenty miles an hour would be enough to send a body out into space forever.

If you did jump off Phobos, however, you would be in no danger of falling down to the enormous disc of Mars, dominating the sky only four thousand miles away. You'd still be moving around Mars in practically the same orbit as Phobos and would complete one circuit in about seven and one half hours. The view would be excellent.

Not many conventional games or sports would be possible with such microscopic gravities. Objects hit or pushed out in the open would dwindle away, swiftly or slowly, on almost flat trajectories, and would never be seen again, except in one peculiar case. Since it is not too difficult to jump off such worlds, it is even easier to establish a satellite orbit around them. In the case of Phobos, the necessary speed is about ten miles an hour. So if you tossed a stone horizontally at exactly the right speed, it would become a moon of Phobos— a satellite of a satellite. Two or three hours later, if your aim had been perfect, it would hit you in the back of the neck.

These examples are enough to prove that anyone who attempts to organize an interplanetary Olympics is going to be in real trouble. A few decades from now, all records listed will have to specify the planet of origin. Some athletic activities may even be laughed out of court; there would be little point in weight lifting, on a world where a ninety-seven ounce weakling could support five tons.

So far we have spoken of sports on worlds that have some gravity, even if it is only a thousandth of the Earth's. But what about the situation in spaceships or space stations, where the very conception of weight— but not mass—is meaningless?

Flying, of course, would be not only easy but unavoidable; it would be the only way of moving around. When we build satellites with really large interior spaces, aerobatics could become an exhilarating recreation, combining the characteristics of ballet and high diving. Zero-gravity athletics is a vast, utterly unknown territory waiting to be explored.

The behavior of liquids in the complete absence of gravity also opens up some interesting possibilities, which belong to the realm of art as much as sport. A weightless liquid forms itself into a sphere under the influence of surface tension. If set vibrating, it will oscillate through all sorts of peculiar shapes. When the oscillations become too large, it may develop a wasp waist and finally fission into smaller drops.

The mobiles of today may be followed by the dynamic water sculpture of tomorrow, which, alas, can never be displayed in any terrestrial art gallery. To see it, patrons will have to go out into space.

Some years ago, I suggested that one of the attractions of a space hotel (and we will have such things in the next century) might be a spherical swimming pool. It might even be a hollow sphere, with an air space inside where spectators might watch their friends swimming around them. I pass lightly over various technical problems—such as that of anchoring the pool in one place, or preventing it from being dispersed by the splashing of the swimmers.

There may be other factors in space besides gravity (or lack of it) which will affect physical activities, and may give rise to new types of sport. It is difficult for us to conceive the possibilities. A dweller in the Sahara, unfamiliar with mountains or beaches, could hardly have imagined ski jumping and surf riding. In the same way we earthlings cannot guess what wholly novel recreations our grandchildren may invent to take advantage of peculiar conditions on the other planets. I will, however, go out on a limb and make a couple of suggestions, but not predictions. They are only mind-stretching hints:

According to some theories, there are lakes and seas on the Moon. But they are not composed of water. They are made of dust, flaked off over millions of years by the expansion and contraction of the lunar rocks, during the 400-degree temperature cycle between day and night. This dust would be so dry, and so finely divided, that it would flow like a liquid and would accumulate in the low lying regions of the Moon. We have nothing quite like it on Earth. Owing to the absence of air it would be slippery and almost frictionless. (If you can imagine the behavior of talcum powder in a vacuum, you have an idea of its characteristics.) Pools of this lunar dust, should they exist, may be a considerable hazard; parts of the Moon may have to be explored on snowshoes, or their equivalent. Yet what opportunities this dust would represent! Think of the frictionless rides one could have on it, in a rocket-driven sledge! Could you ski on it? "Swim" in it? Cruise on it in a boat? We'll soon know.

My second suggestion is much more "far out," but it is based on a number of scientific papers which have appeared during the last few years. These have proposed, in all seriousness, that space travel may best be undertaken by sailing ships.

Few people realize that a great wind continually blows outward from the Sun. It is a wind of light, and it exerts a definite pressure. This "radiation" pressure is, for ordinary purposes, negligible. Tiny though this force may be, it could add up to appreciable amounts over the surface of a huge, gossamer-thin sail of some reflecting material like aluminum foil or mylar film coated with silver. And when I say "huge," I mean exactly that; the sails would have to be thousands of feet across to be of any use. Even so, by using delicate rigging, their mass need be only a few hundred pounds. Once conventional rockets had carried them up into orbit, they could be employed to tow cargoes across space.

Though the acceleration produced by such a "solar sail" is tiny, it would be maintained hour after hour, week after week, and could eventually build up to respectable velocities. The beauty of the system is its utter simplicity, and, above all, the fact that power is free and everlasting. Even if it never has any serious applications, it suggests a beautiful and fascinating sport.

Some day, space yachtsmen will be tacking around the orbit of Mercury, racing tiny one-man vehicles not much larger than the capsules that today's astronauts already have ridden. Billowing ahead of them will be vast, glittering surfaces, possibly miles across—flexible mirrors little thicker than soap bubbles, reefed and furled by a spider's web of invisibly fine threads. The skippers of these fantastic little craft would need a superb knowledge of astronautics and orbital theory, as well as skills that could not be learned in any classroom. There are many links between sea and space; here, surely, is one of the strangest. Across the centuries, the spirit of the men who once sailed the windjammers around the Horn may live again as their descendants ride the eternal trade wind between the worlds.

The haunting vision of these fragile space yachts, literally riding on sunbeams, is sufficient answer to those who think that interplanetary flight will be all cold science and massive engineering. Of course, we shall need that kind of technology to take us to the planets and to build new civilizations there, but this represents only part of life. Our picture of space is not complete if we think of it only in terms of power and knowledge; for it is also a playground whose infinite possibilities we shall not exhaust in all the ages that lie ahead.

Preface 2

The four pieces that follow are all concerned with various aspects of communications satellites, particularly as they will affect human society in the decades to come.

First is an historical note explaining how and when the idea of comsats—a horrid word, but we seem stuck with it—originated. My 1945 paper, though it logically belongs in this section, has been given as an appendix as it is slightly technical and so would be something of a road block to many readers.

"The Social Consequences of the Communications Satellites" is a paper delivered to the symposium on Space Law at the Twelfth International Astronautical Congress, Washington, 1961. "Broadway and the Satellites" was commissioned by Playbill, the remarkable little magazine which appears in multiple editions at all the major New York theaters. The article was written on the (highly improbable) assumption that the playgoer would read it during the interval.

"The World of the Communications Satellite" is a more weighty piece with a somewhat unusual origin and background. In 1963, the International Telecommunication Union—the body which for over a century has regulated the world's electrical, and later radio, communications—called a special conference in Geneva to deal with the explosively expanding subject of space communications. UNESCO, the ITU's parent body, asked me to write a sort of aide-memoire on comsats to provide some guide lines for the conference, and this paper is the result.

The commission was so important that I could not turn it down, but it arrived at an exceedingly awkward time. I received UNESCO's request when living in a shed on the south coast of Ceylon, not far from the equator, surrounded with Aqualungs and underwater cameras and preparing to leave at any moment for a dangerous, wave-swept reef just visible on the horizon ten miles away. On that reef my partner Mike Wilson had discovered, perhaps unwisely, a 250-year-old wreck containing about a ton of silver, and we were rather anxious to get back to it before the news leaked out. This was not a situation conducive to calm, philosophical thought; but, surprisingly enough, it did provide an appropriate starting point for my thesis, which was written in the intervals between the various mechanical and marine disasters recounted in The Treasure of the Great Reef.

Finally: by a coincidence so incredible that it leaves me with a distinct feeling of unreality, I find myself reading these proofs while watching Early Bird inaugurate the first scheduled, transatlantic TV program (May 2, 1965). Thus the passage of almost exactly twenty years has demonstrated to some hundreds of millions of viewers the basic principles I outlined in the last summer of the war.

Now we can proceed to the second item on the agenda—direct spacecasting to home receivers. And that is going to be achieved in very much less than twenty years.

A Short Pre-History of Comsats,
Or: How I Lost a Billion Dollars
in My Spare Time

It is with somewhat mixed feelings that I can claim to have originated one of the most commercially valuable ideas of the twentieth century, and to have sold it for just $40. This cautionary tale begins in May 1945, when I was a flight lieutenant in the Royal Air Force, stationed a few miles from Stratford on Avon. Here I was peacefully engaged—the whole of my war was a very peaceful one—in training airmen to maintain the Ground Controlled Approach radar gear used to talk down aircraft in conditions of bad visibility. Though this work was fascinating (and has since formed the background of my novel *Glide Path*) it left me plenty of time to think about space flight, my chief interest since I joined the British Interplanetary Society in 1935.

Owing to the war, the B.I.S. was in a state of suspended animation, not that this made much difference as far as the rest of the world was concerned. The society had been founded in 1933, and at its peak had achieved a membership of little more than a hundred.[1] (Today, it is several thousand.) A hard core of a dozen enthusiasts kept in touch, by correspondence and occasional private meetings, thoroughout the war. By early 1945 the European conflict was over and peace

[1] See "Memoirs of an Armchair Astronaut," in this volume.

was dimly in sight; so we began to make plans for our postwar activities. I was thus simultaneously immersed in electronics and astronautics; the fact that I was also writing science fiction at the same time probably helped.

As a serving officer, working with highly classified equipment and bound by the Official Secrets Act, I could not submit technical papers for publication until they had been passed by a reviewing board at the Air Ministry. That board had already received two mathematical papers on television waveforms, duly published in *Electronic Engineering* and the *Wireless Engineer,* and now forgotten by myself and everyone else. I would give a good deal to know what it thought when, in July 1945, it received my third epic, which I had boldly entitled "The Future of World Communications." Perhaps they thought it was harmless nonsense; anyway, they gave their approval in a suspiciously short period of time. I then sent the paper to the journal *Wireless World,* which published it in its October 1945 issue under the title "Extraterrestrial Relays."

The proofs arrived a few days after Hiroshima, and I added a perhaps overenthusiastic postscript about the impact of nuclear energy upon astronautics. It actually took nineteen years before the first atomic rocket was run under full thrust, for this was not achieved until 1964.

The paper, which is given in full in the appendix, ran to four pages and four diagrams, and though most of it is quite understandable by those not versed in electronics, I will summarize its main points here in nontechnical terms.

It opened with a short discussion of the problem of long-range radio and TV, pointing out that, for the latter, expensive coaxial microwave cables or relay links would be necessary, and that these could never provide transoceanic services. (In 1945, of course, no such links existed even on *land,* but it was obvious that they would soon be built.) Then, rather gingerly ("Many may consider the solution too farfetched to

be taken very seriously . . .") I introduced the reader
to the idea of artificial satellites, explaining that if a
rocket could reach a speed of five miles a second it
would never fall down, but could continue to circle
the Earth indefinitely like a second Moon. This was a
pretty startling idea for 1945; who could have guessed
that, within fifteen years, the average person would
be unable to remember how many moons the Earth
possessed at any given moment?

The article then pointed out that, although a satellite
could be established at any altitude, so long as it was
clear of the atmosphere, the most interesting and valuable orbit was at a height of twenty-two thousand miles.
At this elevation, a satellite would take exactly one
day to revolve around the Earth; therefore, if it was
placed above the equator, it would appear to stay fixed
in the sky. Unlike all other heavenly bodies, it would
neither rise nor set.

Perhaps I should explain that, so far, there was
nothing original in all this. It might be unfamiliar to
most of my readers, but it was elementary to anyone
interested in space flight. Artificial satellites had been
discussed in the literature of astronautics for almost half
a century, as observatories and refueling bases for outward-bound spacecraft.

"Using material ferried up by rockets," I continued,

it would be possible to construct a "space station"
in such an orbit. The station would be provided
with living quarters, laboratories and everything
needed for the comfort of its crew, who would be
relieved and provisioned by a regular rocket service . . . It could be provided with receiving and
transmitting equipment, and could act as a repeater
to relay transmissions between any two points on
the hemisphere beneath. . . .

Since a single station would only provide coverage
to half the globe, I suggested that three "should be
arranged equidistantly around the Earth, and the fol-

lowing longitudes appear to be suitable: 30 E—Africa and Europe; 150 E—China and Oceania; 90 W—the Americas." Then followed a calculation of the energy required for such a service, in which I arrived at the somewhat optimistic answer that a worldwide FM system would need no more power than the BBC's London TV transmitter.

All the electrical energy needed to run the relay stations could come from the Sun. Except for very short periods round the equinoxes, when they would dip briefly into the shadow of Earth, they would be in continuous daylight, and would intercept a flood of radiation which could be used to operate a heat engine coupled to an electric generator. I also remarked that "thermoelectric and photoelectric developments may make it possible to utilize the solar energy more directly." This is exactly what has happened; the solar cell, invented at the Bell Telephone Laboratories a few years later, now powers almost all satellites and space probes.

This, in brief outline, is the ground covered by my 1945 paper, and subsequent events have confirmed all its main details. Because of rocket payload limitations, the first communications satellites (Score, Courier, Telstar, Relay) were all launched into orbits quite close to the Earth; the first truly stationary or synchronous TV satellite, Syncom 3, was launched on August 19, 1964, just in time for the Tokyo Olympics.

This event, incidentally, is a good example of the perils that beset a prophet. In October 1961, while moderating a panel discussion at the American Rocket Society's "Space Flight Report to the Nation," I had mentioned that the 1964 Olympics would be a good target to shoot for with a synchronous satellite. (I cannot claim credit for the idea, which I'd picked up in general discussions a few days earlier.) Dr. William Pickering, director of the Jet Propulsion Laboratory, was in the front row of my audience, and he was so tickled with the suggestion that *he* passed it on to (then) Vice-President Johnson, speaker at the society's banquet the next evening. The vice-president in turn thought it was such a good idea that he departed from his prepared speech to include it; so when *Profiles of the Future* was published in 1962, I felt confident enough to predict that most large cities would carry live transmissions from Tokyo in 1964.

What I had failed to foresee was that, despite heroic efforts by the White House, the Communications Satellite Corporation, NASA and the Hughes Aircraft Company (builders of Syncom 3) a large part of the United States did *not* see the superb live transmissions from the Olympics which were made available by this triumph of technology. Why? Because they arrived at an awkward time, and the networks did not want to upset their existing program and advertising arrangements!

Another fact which I did not foresee in 1945, though I make no apology for it, was that developments in electronics would make *unmanned* communications satellites possible long before there were any permanent manned space stations. I envisaged my "extraterrestrial relays" as fairly large structures with their own maintenance and operating crews, but miniaturization and, above all, the invention of the transistor, made it possible for tiny robots to do the work of inhabited space stations. Nevertheless, space communications may not be wholly reliable until we can have men on the spot; a troubleshooter who knows how to replace a component costing a few cents can put a multimillion dollar satellite back on the air. There are quite a few dead space vehicles in orbit that could be fixed by a screwdriver and a good mechanic, but are now so much junk costing many times their weight in gold.

During the late forties and early fifties I publicized the idea of communications satellites fairly extensively in books and articles; the synchronous satellite network formed one of the end plates of *The Exploration of Space* (1951), and the Book-of-the-Month-Club edition in 1952 must have introduced the concept to a wide public. I also plugged comsats in my first novel, *Prelude to Space* (1950), and by the mid fifties everyone seriously interested in space travel must have been aware of their potential, though probably few knew where the idea originated.

Can I, indeed, claim that it originated with me, or did I pick up the suggestion from somewhere else? My good friend Dr. John Pierce, director of communications research at the Bell Telephone Laboratories (of whom more anon) has reminded me of a series of stories that appeared in *Astounding Science Fiction* during 1942 and 1943. These tales, written by George O. Smith as a relaxation while developing the radar proximity fuse, concerned the adventures of a group of scientists on an artificial planet known as Venus Equilateral, whose purpose was to serve as a relay station between Earth and Venus when direct radio communication was blocked by the Sun.

Thanks to the generosity of Willy Ley, who kept me supplied with *Astoundings* throughout the war, I had read and enjoyed these stories, and it is quite possible that they had an unconscious influence on my thinking.

Much later, I discovered that Herman Oberth, in his classic 1923 book *The Rocket into Planetary Space,* had suggested that space stations could establish communication with remote places on the Earth by means of mirrors reflecting sunlight. If this seems a little *naïve,* it was because it was written around 1921, when radio was still in its infancy, and the average long-distance station was a mass of wires and antenna towers about a square mile in extent. Since I had never read or even seen Oberth's book at that time, I was quite unaware of the reference.

In any event (until *Pravda* trumps me) I think I can claim priority for the first detailed, specific technical exposition of the global comsat system, with particular reference to synchronous orbits, and the question which frequently nags me is this: *Should* I have published such a multibillion dollar idea in the open literature? Should I have taken any steps to obtain more than the $40 (a fair price at the time) which *Wireless World* paid for the article?

I've no delusions on one point; if I'd not published in 1945, someone else would have done so by 1950 at the very latest. The time was ripe and the concept was inevitable; it was certainly bobbing around in the back of many ingenious minds.

The idea of patenting the concept never occurred to me, and my excuse for this is sheer lack of imagination. Not for a moment did I consider, in that final spring of the war, that the first crude comsat (Score, December 1958) would be orbiting within thirteen years, and that commercial operations would start within twenty.

I now know that, in all probability, I could not have patented the idea in 1945 even if I had made the effort. A lawyer friend, who also tries to earn an honest living writing science fiction, recently looked into the matter and wrote up his conclusions in a story-article "The Lagging Profession" (originally published in *Analog,* January 1951, later reprinted in the *Sixth Annual of the Year's Best SF*)!

As far as I can understand the legal mind, and the labyrinthine intricacies of patent law, I gather from my friend Leonard Lockhard's thesis that (a) I couldn't have patented comsats in 1945; (b) if I had succeeded, the patent would later have been declared invalid, and (c) if it had been valid, it would have been worthless. In support of this argument, so encouraging to all would-be inventors, he quoted the remarkable case of Moffett *v* Fiske.

Admiral Fiske was foolish enough to take out a patent for the torpedo-carrying airplane, back in 1912. The Navy (need I say "Of course"?) would have nothing to do with such a crazy idea, and was a little shaken when the admiral sued many years later and was awarded $198,500. But, alas, the case was reversed on appeal, the higher court arguing that the admiral should not have been granted his 1912 patent because *at that date* there was no airplane capable of carrying a torpedo, and no torpedo capable of standing the shock of being dropped from an airplane. It was useless for the admiral to claim that such developments were a matter of a few years, and that events had fully vindicated him. He didn't collect, and merely proved the truth of the statement that "A patent is nothing more than a license to sue." It is fatal to be too far ahead of your time, and nowadays, "too far" can be about five years.

Even if I had slipped a patent past the examiners in 1945, there is another poignant aspect of the situation. The life of a patent is seventeen years; so it would have expired just as the Communications Satellite Corporation was set up. . . .

Yet I am quite sure that there were all sorts of loopholes that I might have exploited *if* I had been a better businessman and *if* I had realized just how quickly astronautics was going to get off the ground. Perhaps I could have registered a few trademarks and otherwise made an expensive nuisance of myself, to intercept a few of the billions of dollars that will soon be invested in the sky.

But I am not particularly bitter about having missed the boat—or the rocket—and my equanimity is certainly not due to any nobleness of character. For in my heart of hearts, I believe that I've received everything that is due to me in terms of recognition from the people who really matter. My greatest pleasure in this respect was the Franklin Institute's award of its 1963 Stuart Ballantine Medal; this gold medal for developments in communication had previously been awarded to John Pierce, Claude Shannon, and the teams that invented the transistor, the maser, and the laser. In such company, I felt something of an impostor.

For the people who deserve the real credit for communications satellites are those who had to convert my paper plans into hardware that will function flawlessly for months and years on end, thousands of miles above the Earth. I risked nothing except a few hours of my time; but other men have risked their reputations.

It is invidious to mention names in a field where whole armies of scientists and engineers are now working, but I would like to pay tribute to three—with apologies to many others who also deserve credit. First is the ubiquitous John Pierce for his multiple contributions to comsats, recognized in 1964 by the award of the Presidential Science Medal at the White House.

Starting in the early 1950's, John publicized the theory of comsats in technical papers much more detailed and influential than mine. He was also the chief driving force behind the spectacularly successful Echo and Telstar projects; and, with Rudolf Kompfner, he was co-inventor of the traveling-wave-tube which is the wide-band amplifier around which all TV comsats are designed.

Next there is Dr. Harold Rosen of the Hughes Aircraft Company, whose enthusiasm for synchronous satellites resulted in their achievement—in the Syncom series—much earlier than many had imagined possible. And finally there is Robert P. Haviland of the General Electric Company, who has made important studies of comsats, particularly as applied to direct broadcasting, that is, broadcasting from a satellite directly into the home, and not as at present via the local station. It is this type of broadcasting which will make possible the most revolutionary advances in communications, as will be explained in later essays in this book.

These scientists, and their colleagues, are the people who have had to do the real work. The nontechnical layman can have no conception of the gulf between first concept and detailed blueprint—or the sometimes almost equally great gulf between blueprint and hardware. I had all the fun and none of the responsibility, and that is a fair bargain.

Yet sometimes, especially toward the end of the financial year, I like to imagine what I could do with even a minute royalty on the communications satellites' telephone, TV and radio traffic. It would have been pleasant, in my old age, to have gone cruising among the asteroids in my private space-yacht. But there are other and more realistic moments when I think of all the soap operas and laxative commercials that will soon be bathing Earth from pole to pole, and a far different future begins to take shape in my mind.

Perhaps I shall spend the rest of my life trying to prove that communications satellites were invented, not by Arthur C. Clarke, but by another man of the same name.

The Social Consequences of Communications Satellites

In the ability to communicate an unlimited range of ideas lies the chief distinction between man and animal; almost everything that is specifically human arises from this power. Society was unthinkable before the invention of speech, civilization impossible before the invention of writing. Half a millennium ago the mechanization of writing by means of the printing press flooded the world with the ideas and knowledge that triggered the Renaissance; little more than a century ago electrical communication began that conquest of distance which has now brought the poles to within a fifteenth of a second of each other. Radio and television have given us a mastery over time and space so miraculous that it seems virtually complete.

Yet it is far from being so; another revolution, perhaps as far-reaching in its effects as printing and electronics, is now upon us. Its agent is the communications satellite.

It is not necessary to go into technicalities to appreciate why such satellites can transform our communications. Until today, the reliable range of radio has been limited to a few score of miles, for the simple reason that radio waves, like light, travel in straight lines and so cannot bend round the curve of the Earth. The only thing that makes long distance radio possible at all is the existence of the ionosphere, that reflecting layer in the upper atmosphere which bounces back the so-called short waves so that they reach the ground again at great distances from the transmitter. In the process they usually acquire considerable distortion and interference; though they may be adequate for speech, they are almost useless for music, as anyone who has listened to a concert on the short-wave bands knows.

For the still shorter waves, which alone can carry television and other sophisticated types of telecommunication service, the situation is even worse. These are not reflected back from the ionosphere at all, but slice straight through it and out into space. They can be used, therefore, only for what is called line-of-sight transmissions; you cannot (except under freak conditions) pick up a television station from much farther away than you could see it in perfectly clear air. This is why television transmitters, and the microwave relays now used to carry hundreds of simultaneous telephone circuits across the country, are all sited on towers or mountains to obtain maximum range.

Satellites allow the communications engineer to place his equipment, in effect, on the top of a tower hundreds or even thousands of miles high. A single satellite-borne transmitter could broadcast to almost half the Earth, instead of to an area fifty miles in radius; three of these, spaced equally round the equator, could provide any type of communication service between any two points on the globe. This is something that has never before been possible, and it is going to happen within the next few years, for every major firm in the electronics business is now preparing to get into orbit. This is the great Gold Rush of the 1960's, for on the ultrashort radio—and even light—waves which the satellites can flash around the world there is room for millions of television and billions of telephone channels.

What effect will the new types of communications services, and the vastly increased numbers of existing ones, have upon our society and our culture? Before we attempt to answer that, it is worth remembering that it is never possible to foresee the full impact of a major invention, or even of a minor one. Look, for example, at the effect of the humble typewriter, which liberated one half of the human race from centuries of subservience. We males have conveniently forgotten just how few were the occupations—and fewer still the respectable ones—open to women a lifetime ago. Mr. Remington changed all that, and the revolution he wrought was trifling compared with that produced by Henry Ford a little later with the Model T.

Yet communication affects us even more vitally and directly than transportation. A man can live a full and rich life without ever stirring from one spot, so long as he has sufficient channels of information. It is only our age that has made a fetish of rushing around the world; if I remember correctly, it was Aldous Huxley who remarked that speed is the only new vice invented by modern man. Communications satellites, though they may themselves be moving at fifteen thousand miles an hour, may have a remarkably stabilizing influence on the human race. They will abolish a vast amount of the traveling and even of the day-to-day commuting that now seems an unavoidable part of our lives.

For communications satellites will enable us, in effect, to move almost instantaneously to any part of the world. A few figures should be enough to demonstrate this point.

The oceans have always been a major barrier to communications. It required a gigantic effort of technology to provide a telephone cable between Europe and America, carrying only thirty-six voice circuits at a cost of more than a million dollars each. Later cables can carry about a hundred circuits, but there is not much room for further improvement, and it would take ten cables, costing perhaps a hundred million dollars, to provide a single television circuit.

Yet a fairly modest satellite, which we can build to-day, could provide a thousand voice channels across the Atlantic, or alternatively a single television circuit. Looking only a decade or two into the future, one can foresee the time when a network of advanced satellites will bring all points on the Earth into close contact so far as telephony is concerned. It will be as quick and easy to call Australia from Greenland, or South America from China, as it is now to put through a local call. Indeed, by the end of this century all terrestrial calls may be local calls and may be billed at a flat standard rate.

This may have as great an effect on business and social life as the invention of the telephone itself. Just how great that was, we of today have forgotten; perhaps we can remind ourselves by imagining that the telephone was suddenly abolished and we had to conduct all business face to face or else by correspondence carried by stagecoach and sailing ship. To our grandchildren we will still seem in that primitive level of development, and our present patterns of daily commuting a fantastic nightmare. For ask yourself how much traveling you would really have to do if you had an office in your own home equipped with a few simple information-handling machines and wide-screen, full-color television through which you could be in face-to-face contact with anyone on Earth. A good nine-tenths of the traveling that now takes place could be avoided with better communications.

There can be no doubt that satellites will have an especially great effect on the transmission of written and printed information. One idea that has been discussed at some length is the Orbital Post Office, which may make most air mail obsolete in a decade or so. A single satellite, using modern facsimile equipment, could easily handle the whole of today's transatlantic correspondence. Eventually, letters should never take more than a few minutes to be delivered to any point on the Earth, and one can even visualize the time when all correspondence is sent by direct person-to-person facsimile circuits. When that time comes, the post office will cease to handle letters, except where the originals are required, and will concern itself only with parcels.

Another development that will have the most far-reaching consequences is the Orbital Newspaper; this is inevitable once the idea gets around that what most people need is information, not wood pulp. Half a century from now, newspapers as we know them may not exist, except as trains of electronic impulses. When you wish to read the *New York Times,* you will dial the appropriate number on your channel selector, just as today you call a party on the telephone. The front page would then appear on your high-definition screen, at least as sharp and clear as on a microfilm reader; it would remain there until you pressed a button, when it would be replaced by page two, and so on.

Of course, the entire format would be completely redesigned for the new medium; perhaps there would be separate channels for editorials, book reviews, business, news, classified advertising, etc. If you needed a permanent record (and just how often do you save your daily paper?), that could easily be arranged by an attachment like a Polaroid camera or one of the high speed copying devices now found in all modern offices.

Not only the local paper but all the papers of all countries could be viewed in this way, merely by dialing the right number—and back issues, too, since this would require nothing more than appropriate extra coding.

This leads us directly into the enormous and exciting field of information storage and retrieval, which is one of the basic problems of our culture. It is now possible to store any written material or any illustration in electronic form—as, for example, is done every day on video tape. One can thus envisage a Central Library or Memory Bank, which would be a permanent part of the world communications network. Readers and scholars could call for any document, from the Declaration of Independence to the current best seller and see it flashed on their screens.

The Electronic Library is bound to come; its development is being forced by the rising flood of printed matter. Recently, a storage device was announced that could contain everything ever written or printed on stone, paper or papyrus during the last ten thousand years inside a six-foot cube. The problem of encoding and indexing all the world's literature in electronic form so that any part of it can be retrieved and played back is a staggering one, but it has to be solved before our libraries collapse under the weight of their books. And when it is solved, any man on Earth who knows how to dial the right numbers will have immediate access to all printed knowledge, flashed from Central Memory Bank up to the nearest satellite and down again to be displayed on the screen of his receiver. If he wishes, he will be able to store it in his own electronic library for easy reference, as we now record music or conversation on tape, although the recording medium will certainly be much more compact and convenient.

The most glamorous possibility opened up by communications satellites is the one which I originally stressed in 1945—global radio and television. This will be something quite new in the world, and we have no precedents to guide us. For the first time one nation will be able to speak directly to the people of another, and to project images into their homes, with or without the cooperation of the other government concerned. Today's short-wave sound broadcasts are only poor and feeble things compared to those which the clear, interference-free reception from satellites will make possible.

I sometimes wonder if the enormous efforts that most large nations now expend on short-wave broadcasting are worth it, in view of the poor quality of reception. But this will change when the direct and far more efficient line-of-sight services from satellites become available. A Londoner, for example, will be able to tune into NBC or CBS or Radio Moscow as easily and clearly as to the BBC. The engineers and scientists now struggling to establish reliable satellite circuits with the aid of antennas the size of football fields will tell you that this is still years in the future, and they may be right. Nevertheless, most of us will see the day when every home will be fitted with radio and TV equipment that can tune directly to transmitters orbiting thousands of miles above the Earth, and the last barriers to free communications will be down.

Those who are already glutted with entertainment and information from their local stations may be less than enthusiastic about this. However, they are a tiny minority of the human race. Most of the world does not even have radio, still less television. I would suggest, therefore, that though the first use of satellites will be to provide increased facilities between already highly developed countries, their greatest political and cultural influence will be upon backward and even preliterate peoples.

For in the 1970's we will be able to put megawatt transmitters into orbit and will also have reliable battery-powered television receivers that can be mass-produced at a cost which even small African or Asian villages can afford.

Quite apart from its direct visual impact, the effect of TV will be incomparably greater than that of radio because it is so much less dependent upon language. Men can enjoy pictures even when they cannot understand the words that go with them. Moreover, the pictures may encourage them to understand those words. If it is used properly, global television could be the greatest force yet discovered for breaking down the linguistic barriers that prevent communication between men.

Nobody knows how many languages there are in the world; estimates run to as high as six thousand. But a mere seven are spoken by half the human race, and it is interesting to list the percentages. First by a substantial margin comes Mandarin, the language of 15 per cent of mankind. Then comes English, 10 per cent.

After that there is a large gap, and grouped together round the 5 per cent level we find in this order: Hindustani, Spanish, Russian, German, and Japanese. But these are mother tongues, and far more people understand English than normally speak it. On the basis of world comprehension, English undoubtedly leads all other languages.

Few subjects touch upon national pride and prejudices as much as does language, yet everyone recognizes the immense value and importance of a tongue which all educated men can understand. I think that, within a lifetime, communications satellites may give us just that. Unless some synthetic language comes to the fore—which seems improbable—the choice appears to be between Mandarin, English, and, for obvious reasons, Russian, even though it is only fifth on the list and understood by less than 5 per cent of mankind. Perhaps it will be a photo finish, and our grandchildren will be bi- or trilingual. I will venture no predictions, but I would stress again that it is impossible to underestimate the importance of communications satellites in this particular domain.

Television satellites will also present us, and that, soon, with acute problems in international relations. Suppose country A starts transmitting what the government of country B considers to be subversive propaganda. This is happening all the time, of course, but no one complains too bitterly today because the process is relatively ineffective and is confined to radio. Just imagine, however, what Dr. Goebbels could have done with a chain of global TV stations, perhaps capable of putting down stronger signals in many countries than could be produced by the local transmitters, if any.

There would be only two ways of countering such unwanted propaganda. An aggrieved government might try to prevent the sale of receivers that could tune to the offending frequencies, or it might try jamming. Neither policy would be very effective, and jamming could only be carried out from another satellite, which would probably cause protests from the rest of the world, owing to the interference with legitimate transmitters elsewhere.

Though there are obvious dangers and possibilities of friction, on the whole I am very optimistic about this breaking down of national communications barriers, holding to the old-fashioned belief that in the long run right will prevail. I also look forward, with more than a little interest, to the impact of non-commercial television upon audiences which so far have not had much choice in the matter. Millions of Americans have never known the joys of sponsorless radio or television; they are like readers who know only books full of advertisements which they are not allowed to skip. How would reading have fared in these circumstances? And how will Madison Avenue fare, when it no longer controls the video channels? Perhaps the apocalypse of the agencies has already been described in Revelation, chapter 18: ". . . And the merchants of the earth shall weep and mourn . . . for no man buyeth their merchandise any more: The merchandise of gold, and silver, and precious stones . . . and ointments . . . and wine, and oil . . . and chariots . . . and souls of men." This last commodity, I believe, is one expended in massive quantities by commercial television.

The old problem of censorship, over which the law and literature have so often come to grips in dubious battle, will certainly be aggravated when all forms of censorship become impossible. The postmaster general, that traditional guardian of morals, will have no effective control over the ether—nor will anyone else. The possibilities of really uninhibited telecasting from space, if any country was unscrupulous enough to deny normal conventions for the sake of attracting viewers to its channels, are somewhat hair-raising. The crime, bloodshed, and violence for which TV has been so heavily criticized, and the unspeakable "horror comics" that have flooded the Western world in so many millions since the war, show what can happen even in societies that consider themselves enlightened. There will always be people who, to sell their wares or their policies, are willing to appeal to the lowest instincts. They may one day be able to do this across all borders, without hindrance.

But the ether is morally as neutral as the printed page, and on the whole, censorship does more harm than good.

Communications satellites can bring to every home on earth sadism and pornography, vapid parlor games or inflated egos, all-in wrestling or tub-thumping revivalism. Yet they can also expose lies and spread the truth; no dictatorship can build a wall high enough to stop its citizens' listening to the voices from the stars.

These are some of the obvious and predictable effects of communications satellites, but there will be others much more subtle that will have even more profound effects upon the structure of our society. Consider the automobile once again; when it was invented, the assertion was made that it would be useful only in cities—because here alone were there roads on which it could operate. Well, in our efforts to free the automobile from an urban existence, we changed the face of the world and abolished immemorial ways of life. With that analogy in mind, I would like to suggest that the communications satellite may have as great an effect upon time as the automobile has had upon space.

The fact that the world is round and it is thus noon in Washington when it is midnight in Mandalay inconvenienced nobody in the leisurely days before the airplane and the radio. It is different now: most of us have had to take overseas phone calls in the middle of the night or have had our eating and sleeping schedules disrupted by jet transport from one time zone to another. What is inconvenient today will be quite intolerable in ten or twenty years as our communications networks extend to cover the globe. Can you imagine the situation if in your own town a third of your friends and acquaintances were asleep whenever you wanted to contact them? Yet this is a close parallel to what will happen in a world of cheap and instantaneous communications, unless we change the patterns of our lives.

We cannot abolish time zones, unless we beat the Earth into a flat disc like an LP record. But I suggest, in all seriousness, that the advent of global telephony and television will lead to a major attack on the problem of sleep. It has been obvious for a long time that we can't afford to spend twenty years of our lives in unconsciousness, and many people have already stopped doing so. You can now buy a little box that keeps you in such deep slumber, through electronic pulses applied to the temples, that you require only one or two hours of sleep per day.

This suggestion may seem to be fantasy; I believe it barely hints at some of the changes that communications satellites will bring about. What we are building now is the nervous system of mankind, which will link together the whole human race, for better or worse, in a unity which no earlier age could have imagined. The communications network, of which the satellites will be nodal points, will enable the consciousness of our grandchildren to flicker like lightning back and forth across the face of this planet. They will be able to go anywhere and meet anyone at any time, without stirring from their homes. All knowledge will be open to them, all the museums and libraries of the world will be extensions of their living rooms. Marvelous machines, with unlimited information-handling capacity, will be able to speak directly into their minds.

And there's the rub, for the machines can far outpace the capacities of their builders. Already, we are punch-drunk with the news, information, and entertainment that bombard us from a thousand sources. How can we possibly cope with the far greater flood to come, when the whole world—soon, indeed, the whole solar system—will be clamoring for our attention?

There is a Persian legend that warns us of what may come from our efforts to devise a communications system linking all mankind. The story tells of a prince who lost his dearly loved queen and devoted the rest of his life to building a monument that would be worthy of her. He hired the finest craftsmen to raise a palace of marble and alabaster around the sarcophagus; year by year it grew until its towers and minarets became the wonder of the world. Decade after decade he labored, but still perfection eluded him; there was some fundamental flaw in the design.

And then one day, as the prince stood on the gallery above the central aisle of the great mausoleum, he realized what it was that spoiled the perfect harmony. He called the architect and pointed to the now dwarfed sarcophagus that held the queen he had lost so long ago.

"Take that thing away," he said.

So it may be with us. The communications network we are building may be such a technological masterpiece, such a miracle of power and speed and complexity, that it will have no place for man's slow and limited brain. In the end there will be a time when only machines can talk to machines, and we must tiptoe away and leave them to it.

Broadway and the Satellites

Some time before the end of this century, a billion people will attend the same first night. They will do it, of course, not in person, but via the worldwide TV network foreshadowed by the Telstar satellite.

Before we consider the impact of communications satellites on the performing arts, let us see what technical developments may be expected. In a very few years, intercontinental TV will be commonplace. Viewers will think no more of being taken, live, from New York to

New Delhi than they do today of being switched from one town to the next.

At first, however, ordinary viewers will still remain tuned to their local stations: unlike the short-wave ham, the TV viewer will not be able to roam at will over the world. The antennas needed to catch the feeble voices of today's satellites are so enormous, and must be backed by such elaborate receiving and tracking gear, that direct domestic reception is out of the question. We will be able to see only the foreign programs that our local networks decide to relay to us.

But this will not always be the case. Perhaps ten years from now, really high-powered TV transmitters will be put into orbit—and into the very special orbit that allows them to remain fixed in the sky (the so-called synchronous or stationary orbit). When that time comes, you will be able to tune in *directly* to European, Russian or Chinese programs coming straight down from the sky; and there will be weeping and wailing on Madison Avenue.

And on Broadway? Not necessarily, though there will undoubtedly be profound changes, some for the better, some for the worse. Recently I had the interesting experience of seeing, while sitting in one midtown theater, a play being staged in another. The large-screen projection, the presence of the audience, and the usual intervals, gave a sense of immediacy and participation which home TV lacks. One can see a whole new art form springing up, with production techniques similar to those employed in studio TV and the movies, but quite different in other respects. Perhaps we may call it Teletheater.

TT, if it develops, will presumably move into the large number of surplus cinemas now threatened with conversion into bowling alleys or (as in England) legalized gambling dens. Every town or community of any size will have such a theater, and everyone from Alaska to Florida will be able to enjoy an identical performance. The show will be the same; only the audiences will be different. And the expression "Off Broadway" will start to look a little sick.

Now TT can obviously flourish without the use of communications satellites, but they will give it a tremendous boost. When the whole world can tune into the Comédie Française, Drury Lane, the Bolshoi, all the Stratfords, Lincoln Center, La Scala, the incentive to build teletheaters will be very great. And such theaters could, at least in principle, be independent of their own national networks: they could present foreign programs by direct pickup.

It is obvious that home TV, with its own widening frontiers, will compete to some extent with the new theaters; and the recent contest between cinema and TV may be repeated in a slightly different arena. The outcome may be much the same as before; the theaters will provide high definition, color, wide-screen, and stereo sound of a quality which home viewers will be unable to match. The aesthetic effect of all these gimmicks will doubtless be deplorable, at least until the producers learn to handle their new powers with moderation.

Any form of instantaneous, worldwide entertainment raises a whole host of problems—technical, cultural and economic—only a few of which can be mentioned here. The first that comes to mind is the question of language.

This is not as important as might be thought, for it does not affect the majority of the performing arts. Ballet means as much (or as little) to an Eskimo as to a Basque. Even when there are words, the example of opera proves that the audience doesn't have to understand them; only for the play are they vital.

Time, I think, will take care of this difficulty. The development of truly global communications will accelerate, as nothing else has done, the adoption of a world language (almost certainly English) which will be spoken as a second tongue by all educated men. And the revolution in transport which is just getting underway may succeed in making even the English and the Americans bilingual. (I would like to think that the

Peace Corps is the first step toward a world in which *all* men and women spend a couple of their formative years in a foreign country, which they will come to regard as their second home.)

Time zones present more fundamental difficulties. The Late, Late Show is already bad enough, but how many of us would get up in the middle of the night to watch anything short of a world heavyweight fight or the first landing on the Moon? Certainly, few people would enjoy "Hamlet" at 4 A.M.

I suspect (and fear) that this problem will also bring its own solution. When everyone in the world is plugged into the same telephone exchange, and all calls are local ones, sleep is going to become impossible anyway. Physiologists and electronics experts are already at work on a "sleep compressor" which will allow two hours of artificial rest to be as effective as eight of the old-fashioned variety. When this dubious blessing is perfected, well in advance of 1984, the twenty-two-hour day will be upon us. We will need those pre-dawn premieres to occupy our copious leisure time.

And there is the intriguing problem of censorship. For the first time in history, no country will have any control over the entertainment sampled by its citizens. (At least in their own homes: the teletheaters will presumably have to watch their licenses.) Just what may happen if some really uninhibited client gets hold of a channel, I've described in *Tales of Ten Worlds*.

However, I must not raise false hopes. The global communications network will operate according to strict international rules, as all telecommunications have done for the last century. (A big UN-sponsored conference to lay down these regulations took place in 1963.) Nevertheless, there will be a lot of fun when the French beam raw Genet to the United States, and the U.S. aims the riper offerings of T. Williams at the U.S.S.R.

More seriously, it will be of great interest to watch the effects upon standards of performance when productions of all types can be exchanged freely over the whole Earth. Will there be a leveling down, or a leveling up?

When I think that half the globe may soon be receiving instant soap opera from communications satellites, I sometimes feel more than a distant kinship to the late Dr. Frankenstein.

Before you get back for the second act, let me chill your blood a little further. Sooner or later, we are going to learn how to transmit electronically not only sights and sounds, but actual sensations, which will be fed directly into the nervous system. Then you will no longer look at pictures on an illuminated screen: the images will bypass your eyes and you will "experience" them inside your brain. The same thing would be true for sound, touch, smell and all the other senses.

Many years ago, the "Feelies" of Aldous Huxley's *Brave New World* barely hinted at these possibilities, and for a much juicier treatment I refer you to Shepherd ("How To Succeed In Business . . .") Mead's novel *The Big Ball of Wax*. The ultimate development would be a world of dreamers, packed by their millions in the stately pleasure domes of Electro-sensory Entertainment, Inc. Here they would live vicarious lives that might be so much more vivid and exciting than mere reality that they would awake only with the utmost reluctance.

The French aristocrat who remarked haughtily "Life? I leave *that* to my servants," may have spoken for the future. Our descendants a hundred years hence may hand over the running of the world to their robots, while they themselves retire into a realm of electronically induced fantasies. And so they will pass away their lives, wired not merely for sound, but for everything.

I do hope you enjoy the rest of the performance.

The World of the Communications Satellite

This note is being written in a tiny fishing village on the south coast of Ceylon, only a few hundred miles from the equator. There are no telephones, no electric light, no newspapers, no cinemas; there are a few battery-powered radios, but reception is poor on the short-waves and impossible on the broadcast band.

It is difficult for a visitor from one of the more developed countries to imagine the social isolation of such a community—though this village is positively suburban when compared with thousands in the more remote parts of Asia and Africa. Most of the human race exists in a cultural vacuum; it is still divided into myriads of insulated villages or tribes, as it has been since the dawn of history. But now, in a brief moment of time, all this will end. The coming of the communications satellite will make it impossible for any human group—indeed, any individual—to be more than a few milliseconds from any other. The social consequences of this, for good or evil, may be as great as those brought about by the printing press or the internal combustion engine. And they will come upon us much more swiftly.

The progress of astronautics since the opening of the space age only six years ago has been so phenomenal that all the technical problems of the synchronous satellite should be solved by the mid-seventies. Millions need no longer be spent, as they are at present, in the vain quest for absolute component reliability. The communications satellites may not be permanently inhabited, but they will be regularly serviced by low-thrust ferries which can bring repair crews to deal with any emergency within a couple of hours. For by 1975, of course, there will be numerous manned space complexes in orbit, to conduct scientific research and to operate the presently unimaginable zero-gravity, high-vacuum industries of the next generation. The communications satellites of the later part of this century may be part of these systems, and will share their servicing facilities.

With the development of space technology, therefore, all but one of the synchronous satellite's present limitations will pass swiftly into history. Rockets are already under development that can put many tons of payload into the twenty-four-hour orbit; nuclear reactors such as Snap-8 could provide enough power for direct TV transmission to home receivers. Although those concerned with today's satellites have to use the tools currently at hand, they should not let present difficulties and failures condition their thinking. But I certainly do not envy them their problems, for every communications satellite built during the next ten years will be obsolete as soon as it is launched.

The only fundamental objection to the synchronous satellite is the propagation time lag, which of course does not affect radio or TV services, but only telephony. I am convinced that once the inevitability of this time lag is understood and users are taught the proper speaking procedure, they will have no difficulty in handling it. Every generation comes to grips with new techniques; our fathers had to learn to use the dial telephone, our grandfathers, the telephone itself. We are currently grappling with long-distance dialing and ten-figure numbers; our children may think nothing of ending each stretch of conversation with "Over"—and it would probably improve our own telephone procedure, and shorten our talking time, if we did this even now.

In any event, if the delay does prove to be intolerable, then we can establish low-altitude satellites (perhaps in accurately sub-synchronous twelve, eight, six or three-hour orbits) purely for telephony. These might provide a first-class premium rate service; the cheap rate would be through the twenty-four-hour satellites. (As a much longer term possibility, it might be mentioned that there are a number of theoretical ways of achieving a *low-altitude, twenty-four-hour satellite;* but they depend upon technical developments unlikely to occur in this century. I leave their contemplation as "an exercise for the student!")

It seems clear, therefore, that the next ten years will be an interim, transitional period, characterized by small low-powered satellites of the Telstar, Relay and Syncom type. Transmissions from these will be received only by very elaborate and expensive ground tracking stations, which can feed signals into national networks. There will be no question of direct broadcasting to the home viewer or listener; he will still have to rely on his existing local stations, if any, and must accept what they choose to give him.

Even so, the communications satellites of the next few years will have a major impact on world affairs, and particularly upon relations between the United States and Europe. Telstar proved this in the first few weeks of operation, when it established the first TV bridge across the Atlantic. As in the case of the first Atlantic cable a hundred years earlier, contact was intermittent; but—again, as in that case—it will soon become permanent.

Because they share many common interests (including language), already possess extensive rebroadcasting facilities, and are separated by a relatively small time differential, the Americas and western Europe will obviously be the first to benefit from communications satellites. Some of the services which may reasonably be expected, either experimentally or on a full operational basis, during the next decade are:

1. *The Orbital Post Office.* It has been pointed out by S. Metzger that a single 5 mc/s satellite has enough information-handling capacity to carry *all* first-class and airmail correspondence between the United States and Europe. Delivery time would be reduced to approxi-

mately a half, the limit being set by the physical collection and distribution of the mail. One of the chief problems involved in this system is psychological; how would the public react to a postal service in which its letters could be read by unauthorized persons at any point? However, it accepted wartime V-Mail; and for really confidential correspondence, there would be no difficulty in developing private electronic coding systems, so that only the right recipient could receive readable copy.

As ground facilities improve, to keep up with the possibilities offered by the satellites, we may expect high-speed facsimile systems to spread to at least all important towns. For business (as opposed to personal) needs, the distinction between mail, telegram and teletype will rapidly blur. In each case, transmission time will approach zero, with profound effects upon industry and public affairs, and a speeding up of the tempo of life.

2. *Orbital Newspapers*. The simultaneous setting of international editions has already been demonstrated. Influential newspapers such as the London and the New York *Times* will experience a great increase in distribution and immediacy. One of the first countries to benefit from this will be, rather ironically, the United States, which has never possessed a really national newspaper. In the longer run, however, the newspaper as we have known it for the last three hundred years cannot be expected to survive the advent of communications satellites; ultimately, the home presentation will become purely electronic.

3. *Intercontinental Telephony*. As more and more band widths are made available, there will be a tremendous increase in long-distance telephony. It is impossible to set a limit to this; man is a talkative animal, and all estimates that have been made of his need for communication have been swiftly surpassed. Although transatlantic calls may not be substantially cheaper during the next decade, I suspect that well before the end of this same century *all* telephone calls to *anywhere*

may be charged at the same flat rate. (How much of the present cost of a call goes to maintain the equipment that merely calculates the bill?) Eventually, the telephone may become a free public service; it will have the same status as the water supply, for in the society of the future it will be equally essential. Any tax on free communication is detrimental to the public interest.

The full consequences of swift, cheap, and universal person-to-person telephony (and later television) cannot possibly be anticipated at this stage. One can only hint at some trends which may become apparent during the next decade, and dominant during the one that follows. They are:

(a) A great decline in personal correspondence, continuing the trend already started by the telephone—and consequently reducing the need for the "Orbital Post Office" just when it becomes technically possible.

(b) A correspondingly great increase in long-distance personal contacts. The situation will develop over the whole world that now exists in great cities, when intimate friends may talk to each other every day but seldom meet. This would have been totally inconceivable only a hundred years ago.

(c) A steep decline in transportation for other than pleasure purposes. Efficient communications and efficient transportation are, to some extent, opposing influences. If either were perfect (i.e., free and instantaneous) there would be no need for the other. Thus one can envisage the time, in the not-too-distant future, when conferences could be conducted with none of the participants leaving their own countries—or even their homes.

It may be objected that much of the important work at such conferences takes place in private and behind-the-scenes negotiations between individuals, not amenable to telecommunications. But even this will cease to be true as person-to-person services multiply. A generation from now, a delegate in Canberra may be able to locate one in Washington much more easily than either could do so today, if they were both looking for each other among the many committee rooms, private suites, simultaneous lecture sessions, dining rooms and bars typical of a large international conference circa 1960.

Now let us look a little further ahead, to the second phase in the development of communications satellites —particularly the impact of direct broadcasting on the undeveloped countries. This, coupled with the timely invention of the transistor, may give a great stimulus to radio; though we usually associate satellites with television, it must not be forgotten that most areas of this planet cannot yet receive reliable, good quality sound.

However, radio brings us immediately to the question of language. A single orbital transmitter can broadcast high-fidelity sound to half the world—but can it provide programs of equal interest to the Congo Pygmy, the Afghan tribesman, the Greenland Eskimo, the man-about-Manhattan? Obviously not, until they share a common language and, at least to some extent, a common culture.

Communications satellites will make a basic world language mandatory. Just as (in quite recent times) all men had to learn to read so that they could work and even survive in modern society, so in the one world of the very near future they will have to possess a language in common.

There can be little doubt that this language will be English (perhaps Basic English). This has long since ceased to be the property of the English and is now the property of the world—so it can be accepted without reservation even by the newest and most hypersensitive of ex-colonial nations. The alternative of some synthetic language seems improbable; all synthetic languages are born dead, though some pursue a wraith-like existence for a few decades before they are forgotten. A decisive step in favor of English may, indirectly, have already been taken by the mainland Chinese, when they made the decision to adopt the Latin script. It is hard to underestimate the importance of this remarkably courageous and intelligent action, for it opens the road to at least printed (and televised) English for another 15 per cent of the human race.

Though there may be understandable political objections from other nations, justly proud of their literary heritage, one of the first orders of business of UNESCO when direct radio broadcasting from satellites becomes available should be English teaching on a global basis —if possible, over independent UN-owned circuits. Clearly, it would be impossible (and unnecessary) to conduct lessons in each of the six thousand languages now estimated to exist. A mere seven are spoken by half the human race, and if work could start with these, that would be an excellent beginning.

The full potentiality of educational programs from satellites could not be exploited, however, until the arrival of vision; one could hardly teach the written language without this aid (though a great deal could be done with the help of leaflets keyed to the program and distributed in advance). And here I would like to make a suggestion for an interesting compromise between radio and full-scale TV.

It should be possible to develop a cheap and simple slow-scan facsimile-plus-sound receiver which could operate on the normal radio band width, without requiring the approximately thousandfold greater spectrum space needed by TV. Such a device could reproduce line drawings and cartoons (halftones would be unnecessary) at a perfectly adequate speed for educational purposes, where the same picture has to stay in view for a minute or more. It would be the remote equivalent of the teacher's blackboard, and with its aid, even language could be taught to peoples who did not share a single word of their instructor's tongue. It would thus be possible to tape programs suitable for multimillion classes.

The impact of such a device—every element of which lies completely within existing technologies—upon pre-literate societies can perhaps be judged by the following example.

In 1948 Monseñor Jose J. Salcedo, appalled by the poverty and illiteracy in his parish, set up a small radio transmitter in the mountain village of Sutatenze (Colombia). His facilities were quite limited but his objectives were large: reduction of illiteracy and provision of useful information. Starting with a few hours of broadcasting on Saturday evenings to 15 receivers and an audience of some 5,000 listeners, Monseñor Salcedo's programme had grown by 1954 to six hours of daily broadcasting to 16,000 receivers and 200,000 listeners. . . . By now (1960) there are in excess of one million student listeners. . . . With a very modest investment, Monseñor Salcedo has radically transformed rural life over a large part of Colombia. Through communal reception, supported by first-level maintenance by parish priests, he has provided a broadcasting system suited to the meager resources of the people and responsive to their needs.

This gives a foretaste of what may be done by satellite communications to reduce illiteracy and ignorance —*if* we decide to devote them to this service rather than to selling soap. (Not that I am against soap, but I am against pretending that one soap is better than any other, and feel that it is essentially degrading to the communications industry to be dependent upon such impostures.) Because the slow-scan receiver would require a bandwidth of less than 10 kc/s, it avoids the power and frequency allocation problems which blur the global TV picture, and could start functioning in the very near future.

There is no question, however, that global TV will arrive just as soon as it is technically and economically possible. So much has been written on this subject that it is difficult to add anything new to it, but the following comments seem in order.

It is often said that the existence of time zones will inhibit the development of instantaneous, planet-wide communications. This precisely parallels the argument, heard at the beginning of this century, that motorcars would only be of use inside the cities, because, of course, there were no roads elsewhere on which they could operate!

When we have true global communications, our way of life will adapt to them—not vice versa. It would be frustrating to live in a society where at any given time, between a third and a half of one's acquaintances were asleep. This will be the global situation, a quarter of a century from now, and society will have to make some Procrustean adjustments. Current attacks on the problem of sleep may give one answer; perhaps we can condense our present requirements, by electronic means, into an hour or so a day. Or the long range solution, unattractive though it is, may lie in the wholly artificial world where life goes on independently of the Sun, and all the clocks on the planet keep the same time. As will appear later, this solution would not be only unpleasant, but unstable.

The existence of nationally operated, direct-broadcasting TV satellites will immediately focus attention upon two problems which today are merely a minor annoyance, but which tomorrow will be intolerable. They are censorship and jamming. For the advent of communications satellites will mean the end of the present barriers to the free flow of information; no dictatorship can build a wall high enough to stop its citizens listening to the voices from the stars. It would be extremely difficult, if not impossible, to jam satellite broadcasts; any attempt by one country to do so would result in an act of space piracy, or a global telecommunications nuisance which the rest of the world could not permit. One of the immediate objectives of the I.T.U., therefore, should be a total ban on jamming, and a recognition by all nations that such behavior is as uncultured as interrupting other people's conversation by making rude noises. And very much more dangerous, since so many vital lifesaving and navigation facilities now depend upon radio links.

It seems that we have, in the communications satellite, a technical device that may help to enforce good behavior and cooperation even upon reluctant parties. (The meteorological satellite, with which it is closely linked, will do the same.) Its influence will be like that of air transport, though on a much larger scale and affecting whole nations rather than a relatively few favored individuals. The inexorable force of astronomical facts will destroy the political fantasies which have so long fragmented our planet. For when all major artistic productions, entertainments, political and news events can be viewed simultaneously by the whole world, the parochialism and xenophobia of the past will be unable to survive.

This will be one major influence of communications satellites; another, perhaps even more fundamental, may be the reversal of a historic trend which has proceeded with scarcely a break for five thousand years. The traditional role of the city as a meeting-place is coming to end; Megapolis may soon go the way of the dinosaurs it now resembles in so many respects. This century may see the beginnings of a slow but irresistible dispersion and decentralization of mankind—a physical dispersion which will take place, paradoxically enough, at the same time as a cultural unification.

It will be none too soon, for it has been truly said that the measure of man's unhappiness is his estrangement from Nature. There is ample proof of this, in the fact that the most vicious of all savages are now to be found in the rotting stone jungles of our great cities. Civilization, in historic fact as well as in etymology, was the child of the city; but now it has outgrown its parent and must escape from its suffocating embrace.

It will be able to do so, when almost all the sense impressions, skills and facilities that we employ in everyday life become amenable to telecommunications —as they will. For as I concluded in my address[1] to the XIIth International Astronautical Congress in Washington, 1961:

---

[1] "The Social Consequences of Communications Satellites" (this volume).

What we are building now is the nervous system of mankind. . . . The communications network, of which the satellites will be nodal points, will enable the consciousness of our grandchildren to flicker like lightning back and forth across the face of this planet. They will be able to go anywhere and meet anyone, at any time, without stirring from their homes . . . all the museums and libraries of the world will be extensions of their living rooms. . . .

And it will not matter where those living rooms may be; for on this planet, at least, the conquest of space will be complete.

REFERENCES

1. N. I. KORMAN & A. KATZ. "Television Broadcasting from Satellites." Rocket Society Reprinted 2722 American, 1962.
2. S. METZGER. "The Orbital Post Office." A.R.S. Reprint 712, 1958.
3. ARTHUR C. CLARKE. "The Social Consequences of Communications Satellites." *Horizon* (U.S.) January 1962 (in this volume). See also Proceedings of the XIIth International Astronautical Congress, Washington, 1961.
4. ARTHUR C. CLARKE. "Voices from the Sky." Chapter 16, *Profiles of the Future.*
5. D. BEHRMAN. "When the Mountains Move." UNESCO.

Preface 3

This section contains a number of shorter pieces all connected with science in some way, but touching upon astronautics only tangentially or not at all.

First is the speech I made in New Delhi when receiving the 1962 Kalinga Prize from M. René Maheu, director general of UNESCO. The Kalinga Prize, awarded for science writing, is a donation of one thousand pounds presented annually by the Indian industrialist and statesman, Mr. B. Patnaik, and administered by UNESCO. Other winners have been Bertrand Russell, George Gamow, Louis de Broglie, Julian Huxley, and Gerard Piel.

The second piece, "Science and Spirituality," is also a by-product of my Indian visit. It was written for a memorial volume, The Cosmic Conscience, published on the centenary of the birth of the Hindu mystic, Swami Vivekananda.

"H. G. Wells and Science Fiction" is the preface to the Washington Square edition of The Invisible Man and The War of the Worlds, but deals with much wider matters than these two stories. While writing it, I was startled to find that the Colombo public library possessed an autographed set of Wells's works; I shall not forget my emotions at encountering his signature, only slightly faded after forty years in the tropics.

"I am proud to receive the Kalinga Prize, an honor which I have coveted ever since it was founded." Those words were spoken last year by my distinguished colleague and compatriot, Professor Ritchie Calder, and they express my own sentiments so perfectly that I cannot do better than repeat them.

I would also like to thank the generous donor of the prize, Mr. Patnaik, and the UNESCO officials who have organized this meeting. It is my hope that, as the years pass, the great importance of this award will become universally recognized and its fame ever more widespread.

In addition to the pride I personally feel on receiving the Kalinga Prize, I would like to think that it is a tribute to the field of literature in which I have specialized—science fiction. Although at least four of the earlier prize winners have written some science fiction, it has been only a minute and incidental portion of their output. I can claim that it is a major part of mine, for I have published just about as much fiction as nonfiction.

Many scientists, I am sorry to say, still look down on science fiction and lose no opportunity of criticizing it. For example, they often point out that 90 percent of science fiction is rubbish—ignoring the fact that 90 per cent of *all* fiction is rubbish. Indeed, I would claim that the percentage of competent writing in the science-fiction field is probably higher than in any other. This is because much of it is a labor of love, written by enthusiasts who have considerable scientific knowledge and who are often themselves practicing scientists.

What role does science fiction actually play in the popularization of science? Though it often serves to impart information, I think its chief value is *inspirational* rather than educational. How many young people have had the wonders of the universe first opened up to them, or have been turned to a scientific career, by the novels of Verne and Wells? Many distinguished scientists have paid tribute to the influence of these great masters, and a careful survey would, I believe, reveal that science fiction is a major factor in launching many youngsters on a scientific career.

It is obvious that science fiction should be technically accurate, and there is no excuse for erroneous information when the true facts are available. Yet accuracy should not be too much of a fetish, for it is often the spirit rather than the letter that counts. Thus Verne's *From the Earth to the Moon* and *A Journey to the Center of the Earth* are still enjoyable, not only because Verne was a first-rate storyteller, but because he was imbued with the excitement of science and could communicate this to his readers. That many of his "facts" and most of his theories are now known to be incorrect is not a fatal flaw, for his books still arouse the sense of wonder.

It is this sense of wonder that motivates all true scientists, and all true artists. We encounter it in the writings of such scientific expositors as Fabre, Flammarion, Jeans, Rachel Carson, Loren Eisley, as well as many of my precursors at this function; and we meet it again in all scientific romances that are worthy of the name. Any man who can read the opening pages of Wells's *The War of the Worlds* or the closing ones of *The Time Machine* without a tingling of the blood is fit only for "treasons, stratagems and spoils."

The cultural impact of science fiction has never been properly recognized, and the time is long overdue for an authoritative study of its history and development. Perhaps this is a project that UNESCO could sponsor, for it is obvious that no single scholar will have the necessary qualifications for the task. In one field in particular—that of astronautics—the influence of science fiction has been enormous. The four greatest pioneers of space flight—Tsiolkovsky, Oberth, Goddard, and Von Braun *all* wrote science fiction to propagate their ideas (though they did not always get it published!).

In spreading the ideas of space flight, science fiction has undoubtedly helped to change the world. More generally, it helps us to face the strange realities of the universe in which we live. This is well put in an article recently sent to me by a science-fiction fan who also happens to be a Nobel Prize winner, Dr. Hermann J. Muller, whose discovery of the genetic effects of radiation has inadvertently inspired much recent science fiction and made "mutant" a modern bogey word. To quote Dr. Muller ("Science Fiction as an Escape," *The Humanist,* 1957, No. 6):

> The real world is increasingly seen to be, not the tidy little garden of our race's childhood, but the extraordinary, extravagant universe descried by the eye of science. . . . If our art . . . does not explore the relations and contingencies implicit in the greater world into which we are forcing our way, and does not reflect the hopes and fears based on these appraisals, then that art is a dead pretense. . . . But man will not live without art. In a scientific age he will therefore have science-fiction.

In the same paper, Dr. Muller points out another valuable service that this type of literature has performed:

> Recent science-fiction [he writes] must be accorded high credit for being one of the most active forces in support of equal opportunities, goodwill and cooperation among all human beings, regardless of their racial and national origins. Its writers have

been practically unanimous in their adherence to the ideal of "one free world."

That, I think, is inevitable. Anyone who reads this form of literature must quickly realize the absurdity of mankind's present tribal divisions. Science fiction encourages the cosmic viewpoint; perhaps this is why it is not popular among those literary pundits who have never *quite* accepted the Copernican revolution, nor grown used to the idea that man may not be the highest form of life in the universe. The sooner such people complete their education, and re-orientate themselves to the astronomical realities, the better. And science fiction is one of the most effective tools for this urgent job.

For it is, preeminently, the literature of *change*—and change is the only thing of which we can be certain today, thanks to the continuing and accelerating scientific revolution. What we science-fiction writers call "mainstream literature" usually paints a static picture of society, presenting, as it were, a snapshot of it, frozen at one moment in time. Science fiction, on the other hand, assumes that the future will be profoundly different from the past—though it does not, as is often imagined, attempt to *predict* that future in detail. Such a feat is impossible, and the occasional direct hits of Wells and other writers are the result of luck as much as judgment.

But by mapping out *possible* futures, as well as a good many impossible ones, the science-fiction writer can do a great service to the community. He encourages in his readers flexibility of mind, readiness to accept and even welcome change—in one word, *adaptability*. Perhaps no attribute is more important in this age. The dinosaurs disappeared because they could not adapt to their changing environment. We shall disappear if we cannot adapt to an environment which now contains spaceships and thermonuclear weapons.

Sir Charles Snow ends his famous essay "Science and

Government" by stressing the vital importance of "the gift of foresight." He points out that men often have wisdom without possessing foresight. Perhaps we science-fiction writers sometimes show foresight without wisdom; but at least we undoubtedly *do* have foresight, and it may rub off onto the community at large.

Before concluding, I would like to take this unique occasion of the first Kalinga presentation on Indian soil, to speak about the promotion of the scientific outlook in the East. Though this task is important enough in the West, it is even more desperately urgent here. Two of the greatest evils which afflict Asia, and keep millions in a state of physical, mental, and spiritual poverty, are fanaticism and superstition. Science, in its cultural as well as its technological sense, is the great enemy of both; it can provide the only weapons that will overcome them and lead whole nations to a better life.

For fanaticism is incompatible with the open-minded, inquiring spirit of science—with the readiness to accept the discipline of external reality, even if it conflicts with one's personal hopes and beliefs. The motto of the fanatic is "Don't confuse me with the facts—I've made up my mind." This is the exact antithesis of the scientific outlook.

As for superstition—most of us can remember, though too many people have already forgotten, the events of last February 5 (1962). On that date a natural and inevitable grouping of the planets (that has happened about twenty times since the days of the Kalinga empire!) caused needless fear to millions. How many lakhs[1] of rupees were then expended to ward off astral influences? And most of that money was spent by families who could ill afford it.

That was a spectacular example of the evils of superstition, but there are countless others unnoticed by the world. Recently, not far from my home in Ceylon, a villager was bitten by a snake. He could get no medical

---

[1] This useful unit, not available in the West, is 100,000.

treatment, because the date was inauspicious; and so he died.

Gentlemen, two years ago Monsieur Jean Rostand, at this very function, referred to your country as "that great nation which welcomes the future without rejecting the past." That is a good policy for any nation—as long as it realizes that there are things in the past that *must* be rejected. Science, which after all is only common sense raised to the nth degree, can tell us what to preserve and what to reject. Heed its voice—if not for your own sakes, then for the sake of the lovely, dark-eyed children of Asia and Africa, who are born in millions every year—and die in millions the next. Their only hope of a better future lies in science combined with wisdom and foresight. I shall be happy indeed if any writings of mine have helped toward this goal.

Memoirs of an Armchair Astronaut (Retired)

For my money, the heroic period of the space age lay between 1935 and 1955; what's happened since has had a slight air of anticlimax. True, men are now actually preparing to go to the Moon, but today everyone takes a little thing like that for granted, and eminent scientists no longer rise in their wrath to denounce rocketeers as irresponsible crackpots. The only arguments about space that one hears today are of this type: Should Brobdingnag Astrodynamics or Consolidated Aerospace be awarded the $326,709,163 contract for the first stage Mastodon booster?

It was all very different in the prewar years, when the annual income of the British Interplanetary Society was about $300. (I should know; as treasurer, I had the terrifying responsibility of accounting for it.) On the other side of the Atlantic, the American Rocket Society was slightly more affluent, but as we both operated with a part-time, volunteer secretarial staff, contact between our two organizations was erratic. In those days, moreover, B.I.S. and the A.R.S. were divided by an ideological gulf, long since bridged.

As is well known, we British are a romantic and wildly imaginative race, and to our annoyance the conservative Americans did not consider that space travel was respectable. Though they had formed the American *Interplanetary* Society in 1930, the name had been changed to American *Rocket* Society a few years later. The suggestion was sometimes made that we should follow suit, but we refused to lower our sights. To us, the rocket was merely the interplanetary bus; if a better one came along (it hasn't yet, but we're still hoping) we would transfer, and give the rocket back to the fireworks industry.

Picture us then, in the mid thirties, when only a few aircraft had flown at the staggering speed of three hundred miles an hour, trying to convince a skeptical world that men would one day travel to the Moon. There were about ten of us in the hard core of the society, and we met at least once a week in cafés, pubs, or each others' modest apartments. We were almost all in our twenties, and our occupations ranged from aeronautical engineer to civil servant, from university student to stock exchange clerk. Few of us had technical or scientific educations, but what we lacked in knowledge we made up in imagination and enthusiasm. It was, I might add, just as well that we were overoptimistic. If we had even dreamed that the price of the first round trip ticket to the Moon would be $10 billion per passenger, and that spaceships would cost many times their weight in gold, we should have been much too discouraged to continue our quarter-million-mile uphill struggle.

The total amount spent on the British space effort before the outbreak of war was less than $1,000. What did we do with all that money? Let me tell you.

Most of us talked, some of us calculated, and a few of us drew—all to considerable effect. Slowly there emerged the concept of a space vehicle which could carry three men to the Moon and bring them back to Earth. It had, even for a 1938 spaceship, a number of unconventional features, though most of them are commonplace today, and many have been "rediscovered" by later workers. Notable was the assumed use of solid propellants, of the type now employed in Polaris and similar missiles. Our first plans, based on highly unrealistic assumptions, envisaged making the entire round trip in a single vehicle, whose initial weight we hopefully calculated at about a thousand tons. (The advanced Saturns now being developed by NASA weigh several times as much.) Later, we discussed many types of rendezvous and space-refueling techniques, to break down the journey into manageable stages. One of these involved the use of a specialized "ferry" craft to make the actual lunar landing, while the main vehicle remained in orbit. This, of course, is the approach now being used in the Apollo Project—and I am a little tired of hearing it described as a new discovery. For that matter, I doubt if *we* thought of it first; it is more than likely that the German or Russian theoreticians had worked it out years before.

There is a vast gulf, almost unimaginable to the layman, between thinking of an idea, and then converting it into detailed engineering blueprints. There is an equally great gulf between the blueprints and the final hardware, so we cannot claim too much credit for our pioneering insight. Yet I am often struck by the fact that there is hardly a single new conception in the whole field of current space research; everything that is happening now was described, at least in outline, twenty or even fifty years ago.

But back to our Model T. As soon as we had finished the drawings, we published them in the minute *Journal of the British Interplanetary Society*. It took us some time to collect enough money to pay the printer; he was a Greek, I remember, and a few Hellenic spellings slipped through my proofreading. Nor am I likely to forget the day when I collected the entire edition, in two parcels, and was walking home with it to the apartment I shared with another space enthusiast a few blocks east of the British Museum. I had got halfway when two polite gentlemen in mackintoshes tapped me on the shoulder, and said, "Excuse me, sir, but we're from Scotland Yard. Could we see what you have in those packets?"

It was a reasonable request, for at the time wild Irishmen were blowing up post offices to draw attention to their grievances, and the Yard was trying to round them up. (They did catch a brat named Brendan Behan, I believe.) To the considerable disappoinment of the detectives, I was not even carrying *Tropic of Capricorn,* but when I presented them with copies of the *Journal* they very gamely offered to pay. Tempting though it was to acquire a genuine subscriber (the cash box held about $2.50 at the time), I refused the contribution; but I got them to carry the parcels the rest of the way for me.

The *Journal* attracted a surprising amount of attention and a not surprising amount of amusement. That *doyen* of scientific publications, the good, gray *Nature,* condescended to notice our existence, but concluded its review with the unkind cut: "While the ratio of theorizing to practical experimentation is so high, little attention will be paid to the activities of the British Interplanetary Society."

That was a quite understandable comment, but what could we do about it, with that $2.50 in the till? Why, launch an appeal for an Experimental Fund.

We did so, and the money came rolling in. There was one occasion, I now blush to recall, when I shared sardines on toast with an elderly lady member in an

Oxford Street tearoom and convinced her that, for $60, one could solve the basic problems of building a meteorological rocket. Eventually we rounded up a couple of hundred bucks, and the research program was under way. (At Peenemünde, though we were not to know it for quite a while, von Braun was already heading for his first hundred million.)

All this money was something of a responsibility; having appealed for it, we had to use it, in a manner most calculated to produce both scientific results and publicity. The actual building and launching of rockets was frowned upon, for it would only result in police proceedings under the 1875 Explosives Act, as a group of experimenters in the north country had already proved.

We were in the position of someone who couldn't afford a car, but had enough for the speedometer and the rear view mirror. This analogy is quite exact; though we couldn't make a down payment on even a compact spaceship, we felt we could develop two of the instruments needed to operate it.

It was a sensible decision, and indeed about the only one possible in the circumstances. The first project we tackled was a spaceship speedometer which had been invented by Jack Edwards, the eccentric genius who headed our research effort.

Edwards, who is now dead, was a short, bearded, and excitable Welshman—and the nearest thing to a mad scientist I have ever met outside fiction. He was the director of a very small electronics firm, which soon afterward expired thanks to his assistance; but he had an altogether uncanny grasp of the principles of astronautics. He had invented, back in 1938, what is now called inertial guidance—the technique which allows a rocket to know just where it is, and how fast it is going, by continually keeping track of the accelerations acting upon it.

Edwards's space speedometer consisted of a large aluminum disk, pivoted on ball bearings, and with sundry gears, weights, and springs attached to it. As the device was moved up or down, the weights would "sense" the forces acting upon them, and the rotation of the disk would record the distance moved. We had planned to test the gadget on one of the deeper elevators of the London Underground, but, you will not be surprised to learn, it never got as far as that. The theory of the device was perfectly sound, and something similar steers every satellite into orbit today. But the engineering precision demanded was utterly beyond our means, and Mrs. Edwards put her foot down on hearing of our intention to cast lead weights in her best saucepan.

Balked on the speedometer front, we tried our luck with the rear-view mirror. To keep it on course during take off, and to provide the crew with artificial gravity, we had proposed to spin our spaceship like a rifle bullet. (The spin would be imparted by water jets, as the ship floated in a kind of raft before launching.) Even though the rate of rotation was quite low, it would obviously be impossible to take observations of the stars from our cosmic carousel, so we had to invent an optical system to unscramble the ship's spin.

This required no great originality, for the astronomers (who also looked out at the stars from a spinning vehicle, the planet Earth) had solved the problem years before. Their answer is an instrument called a coelostat, which, however, has to cope with only one revolution every twenty-four hours. We built a similar arrangement of four mirrors—two fixed, two spinning —and I sacrificed the spring motor of my phonograph to provide the motive power.

The coelostat *worked;* it was the only thing we ever made that did. Its public demonstration took place in most auspicious surroundings, the hallowed halls of the South Kensington Science Museum, whose director deserves much credit for providing hospitality to such a far-out organization as ours. Next to the room where we held our meeting was the original Wright biplane, still in exile from the United States; on the floor above was an even more momentous piece of machinery—the "atom smasher" with which Cockroft and Walton had produced the first artificial nuclear reaction in 1932.

Our setup was simple, but effective. At one side of the room was a disk with lettering on it, spinning too rapidly for the words to be read. At the other was the coelostat—a wooden box about a foot on a side, looking rather like the result of a mésalliance between a periscope and an alarm clock. When you peered through the coelostat at the spinning disk, the latter appeared to be quite stationary and you could read the inscription "B.I.S." painted on it. If you looked at the rest of the room, however, it appeared to be revolving rapidly; this was not recommended for any length of time.

Though our experimental efforts were unimpressive, we made ourselves known through countless lectures, newspaper interviews, and argumentative letters to any publications that would grant us hospitality. One controversy ran for months in the correspondence columns of the BBC's weekly, *The Listener;* if we could not convince our critics, we usually routed them.

Looking back on it, I am amazed at the half-baked logic that was used to attack the idea of space flight; even scientists who should have known better employed completely fallacious arguments to dispose of us. They were so certain that we were talking nonsense that they couldn't be bothered to waste sound criticism on our ideas.

My favorite example of this is a paper which an eminent chemist presented to the British Association for the Advancement of Science. He calculated the energy that a rocket would need to escape from the Earth, made a schoolboy howler in the second line, and concluded, "Hence the proposition appears to be basically impossible." But that was not enough; he could not resist adding, "This foolish idea of shooting at the Moon is an example of the absurd lengths to which vicious specialization will carry scientists working in thought-tight compartments." I cannot help feeling that the good professor's compartment was not merely thought-tight; it was thought-proof.

As another example of the sort of stick that was used to beat us, I might mention an article that appeared under the eye-catching title "We Are Prisoners of Fire." This was based on the fact, deduced from radio measurements, that there are layers in the upper atmosphere where the temperature reaches a couple of thousand degrees Fahrenheit. Therefore, the writer announced, any space vehicle would melt before it got more than a few hundred miles from Earth. He had overlooked the point that, at the altitudes concerned, the air is so tenuous that the normal concept of temperature has no meaning, and one could freeze to death for all the heat that the few 2,000-degree molecules of nitrogen and oxygen could provide.

I must admit that we thoroughly enjoyed our paper battles. We knew that we were riding the wave of the future; as T. E. Lawrence said in *Seven Pillars of Wisdom,* "It felt like morning, and the freshness of the world-to-be intoxicated us." But the world-to-be was moving inexorably, unmistakably toward war. I remember sending out, from the fourth floor apartment in Gray's Inn Road that was both my residence and the B.I.S. headquarters, an emotional farewell to all our hundred members, and then descending to the shelters as the sirens gave their warning.

But it was a false alarm; nothing happened then, or for a long time afterwards. Finding to our surprise that we had not all been blown to pieces, we resumed contact and continued our discussions, by means of correspondence and occasional private meetings. As an R.A.F. instructor, I was in a position to indoctrinate hundreds of hapless airmen, and made the most of the opportunity. For some odd reason, my service nickname was "Spaceship."

At last it was winter 1944. The European conflict was clearly drawing to an end—but though there was nothing about it in the papers, for several weeks large holes had been suddenly appearing in southern England. Despite this, we were holding a meeting in London to plan our postwar activities. The speaker had just returned from a mission in the United States, where a well-known authority had assured him that tales of large German war rockets were pure propaganda. We were still laughing at this when—CRASH!—the building shook slightly, and we heard that curious, unmistakable rumble of an explosion climbing backward up the sky, from an object that had arrived faster than the sound of its own passage . . . A few months later, when we knew his address, we hastened to confer the honorary fellowship of the society on Dr. Wernher von Braun.

The post V-2 world, of course, took us much more seriously. Few people now doubted that rockets could travel great distances into space, and most were prepared to admit that men could travel with them. We had to alter our propaganda line; it was no longer necessary to spend all our efforts proving that space flight was possible—now we had to demonstrate that it was desirable. Not everyone agreed with us.

One who did was George Bernard Shaw, who joined the Society in his ninety-first year and remained a member until his death. He was a personal capture of whom I was very proud; in 1946, while still at college, I sent him a copy of my philosophical, Toynbee-inspired paper "The Challenge of the Spaceship." To my surprise, back came one of the famous pink postcards, followed soon afterwards by a longer communication containing some typically Shavian theories of transsonic flight. If you are interested, you will find the whole of the brief Shaw-Clarke correspondence in *The Virginia Quarterly Review* for Winter 1960.

Less sympathetic to our aims was Dr. C. S. Lewis, author of two of the very few works of space fiction that can be classed as literature, *Out of the Silent Planet* and *Perelandra*. Both of these fine books contained attacks on scientists in general, and astronauts in particular, which aroused my ire. I was especially incensed by a passage in *Perelandra* referring to "little rocket societies" that hoped to spread the crimes of mankind to other planets. And at the words: "The destruction or enslavement of other species in the universe, if such there are, is to these minds a welcome corollary," I really saw red.

An extensive correspondence[1] with Dr. Lewis led to a meeting in a famous Oxford pub, the Eastgate. Seconding me was my friend, Val Cleaver, a space buff from way back (and now chief engineer of the Rolls-Royce Rocket Division). Supporting Lewis was Professor J. R. R. Tolkien, whose trilogy *The Lord of the Ring* created a considerable stir a few years ago. Needless to say, neither side converted the other, and we refused to abandon our diabolical schemes of interplanetary conquest. But a fine time was had by all, and when, some hours later, we emerged a little unsteadily from the Eastgate, Dr. Lewis's parting words were "I'm sure you're very wicked people—but how dull it would be if everyone was good."

The postwar years brought a new and novel problem which is still with us, though in less virulent form. From 1948 onward, as you will doubtless recall, the sky started to fill with flying crockery; there were times when hardly a day went by without press reports of visitors from space. We were, obviously, the last people to deny this possibility; but we were quite sure that the arrival of genuine spaceships from elsewhere could no more be kept secret than the presence of a hungry Tyrannosaurus Rex in St. James's Park.

The Flying Saucers caused us considerable embarrassment and annoyance, because there was a danger that in the public eye we should be associated with the cranks and crackpots who were spearheading the cult. In an attempt to strike a blow for sanity, I did a half-hour TV program exposing a gentleman who claimed to have made contact with Saucerites. My quest for ammunition led me to a second meeting with Scotland Yard, whose photographic experts examined the crudely faked "evidence" and gave me some useful unofficial

---

[1] This will appear in the volume of the late Dr. Lewis's correspondence now being edited by his brother Major Lewis.

advice. I promptly returned to my own darkroom and produced a much better set of flying saucers, which proved conclusively that (a) any number can play and (b) the original photographer had been careless, because some of his saucers were clearly *inside* his telescope tube.

Though the society still had no money, it was a good deal larger than in the prewar days, and the quality of its membership considerably more impressive. Our bimonthly *Journal* was widely read; in particular, the Soviet embassy subscribed to twenty copies. And here is a very odd thing; though the Russians purchased the *Journal* in bulk and arranged their own distribution, they sent us a complete list of all the scientific and technical institutions in the U.S.S.R. which received copies. We would never have dreamed of asking for such a document, but it arrived unsolicited and made fascinating reading. I passed it on to the parties who should have been interested; as it turned out, they apparently weren't.

With growing maturity, and a better understanding of what still remained to be done in the way of engineering development, we decided to concentrate on nearer objectives than the Moon and planets. By the late 1940's it was obvious that small satellite vehicles could be developed in the near future, and would be of enormous scientific value.

In 1951, all these ideas came to a head when we arranged an international congress in London, on the theme of the artificial satellite. It was well attended by scientists from many countries, and one paper described the construction of a satellite vehicle of a size and performance very similar to the later Vanguard. This was designed to put into orbit an inflatable metallized balloon; less than ten years later, the whole world was to watch such an object—the moving star of Echo I.

By this time (and, if I may say so, none too soon) official circles in the United States had started to take a mild interest in space. A few farsighted individuals had already done much more, frequently to the annoyance of their superiors. (I once heard General Shriever remark that he still keeps, in his safe, a Department of Defense directive forbidding him to use the word "space" in any public statements.)

Among the postwar American converts was a young physicist named Fred Singer, then a science attaché with the U.S. Office of Naval Research. He had already done notable work with rocket probes in the upper atmosphere, but was somewhat skeptical about space flight. However, after a few brainwashing sessions, he became wildly enthusiastic, and we soon had to hold him down lest he start galloping all over the solar system.

One evening Fred, Val Cleaver, and I were sitting in the Arts Theatre Club, thinking of ways to drum up interest in scientific satellites. "What we want," said someone, after the second or third round of drinks, "is a nice snappy name for the project." That started us doodling, and after a little while we concocted the abbreviation MOUSE, for "Minimum Orbital Unmanned Satellite of Earth." In the next few months, Fred produced a blizzard of papers describing what MOUSE (better still, MICE) could do; his predictions were uncannily accurate, and every one of them has since come true. The publicity campaign was extremely successful, and MOUSE appeared in technical journals all over the world. Indeed, a few years later, an American news agency picked up one of Fred's drawings from a Russian paper and hawked it around as an example of a genuine Sputnik!

Our conversion of Fred Singer into a space cadet was probably one of the most important things we ever did. Quite apart from his missionary work in the more backward and savage areas of United States science, he played a dominant role on the committee that recommended the launching of an IGY satellite. Though several other groups, and many individuals, were working toward the same end, Singer's intervention at a crucial moment, known only to a handful of people, was quite possibly decisive in committing the United

States to a satellite program. That it was the wrong satellite was not his fault.

On July 29, 1955 I was about as far from Washington as one could get, for I was living in a small wooden hut on an island of the Great Barrier Reef, thirty miles off the coast of Queensland. Coming in from a morning's diving along the reef, I happened to switch on the radio and was transfixed by the news that President Eisenhower had authorized the launching of scientific satellites during the International Geophysical Year. At great difficulty and expense I dispatched Singer a cable saying "Congratulations—may MOUSE bring forth a mountain." When the message finally reached civilization by pearling lugger and aborigine postman, it was indecipherably distorted, and to add insult to injury the charges had got reversed in the process. It was a couple of years before I caught up with Fred and straightened things out.

Just as the V-2, in 1945, marked the end of the first era of astronautics, so the announcement of Project Vanguard, ten years later, marked the end of the second. As far as we old space hands were concerned, the long campaign was over. A major power was now in the satellite business, reluctantly but inescapably. Given time, everything that we had predicted was bound to follow. Some of us hoped that we might live to see the first landing on the Moon—though in one of my early novels I had stuck my neck out by suggesting 1978 as a target date. Today, anyone so pessimistic would be extremely unpopular at NASA headquarters.

That our time scale might be a little inaccurate I began to suspect in the small hours of October 4, 1957, when a London paper roused me from my bed in a Barcelona hotel and asked if I cared to comment on a news flash just received from Moscow. There is no need to elaborate upon what has happened since then; it is enough to list some of the names that have now passed into history: Sputnik, Laika, Lunik, Gagarin, Shepard, Titov, Glenn, Mercury, Telstar,

Mariner . . . these are merely the first words in the vast new vocabulary of space.

It has been a privilege to watch the beginnings, and to have taken some small part in the greatest adventure upon which the human race has ever embarked; but now it has grown too unimaginably huge for the comfort of amateurs like myself. This has struck me many times in the last few years—never so strongly as in the Grand Ballroom of the Waldorf Astoria, in the fall of 1961.

There, some two thousand scientists and engineers, all in evening dress, had assembled for the banquet which concluded the American Rocket Society's Space Flight Report to the Nation. The cream of the astronautics industry (soon to be the largest business in the world) was gathered together; had the roof fallen in, that would have been the end of the United States' space effort, and of its Vice-President Johnson, for he was the guest of honor, speaking on a nationwide hookup.

Yes, it was an impressive occasion, and I was happy to be there. But I could not help thinking of the little pubs and tearooms where we met between the wars, and dreamed the dreams we never thought to see come true.

The new generation will know the drama, the triumphs, the excitement, the responsibility of space flight.

But we had most of the fun.

Science and Spirituality

A few days ago I heard Sri Nehru quote a striking remark of Vinobe Bhave: "Politics and religion are obsolete; the time has come for science and spirituality." Not the least reason why the phrase is striking is that, to many people, science and spirituality are not merely incompatible; they are antagonistic.

It is a great tragedy that such an impression has ever arisen, for nothing could be further from the truth. "Truth"—that, of course, is the key word; for what does science mean except truth? And of all human activities, the quest for truth is the most noble, the most disinterested, the most spiritual.

It is also the one most liable to inculcate humility. Said T. H. Huxley a century ago: "Sit down before fact as a little child, be prepared to give up every preconceived notion, follow humbly wherever and to whatever abysses nature leads, or you shall learn nothing."

Science has now led us, in our generation, to the ultimate abyss—that of space. Questions to which philosophers and mystics have given conflicting answers for millennia will soon be answered, as our rocket-borne instruments range ever further from Earth.

Of all these questions, the place of intelligence in this gigantic universe of a hundred thousand million suns is the most important, the one that most teases the mind. During the last decade, the idea that life was a very rare and peculiar phenomenon, perhaps existing only upon our planet, has been completely demolished.

All the evidence today suggests that it is commonplace; within ten years we may know.

But even if we encounter life on the other planets of this Sun, it seems most unlikely that we shall meet intelligence. The odds are fantastically against it; since the solar system is at least five thousand million years old, it is altogether unreasonable to expect that other rational beings will be sharing it with us at this very moment.

To find our peers, or more likely our superiors, we must look to the stars. There are still some conservative scientists (it is astonishing how often people fail to learn from the past!) who would deny that we can ever hope to span the interstellar gulfs which light itself takes years to cross.

This is nonsense. In the foreseeable future (it may take a century, but what is that?) we shall be able to build robot explorers that can head to the stars, as our present ones are heading to Mars and Venus. They will take years upon their journeys, but sooner or later one will bring back news that we are not alone.

That news may also reach us, more swiftly and in richer detail, in the form of radio or other messages. On our own planet, radio techniques have been in existence for the span of a single lifetime, yet within that short interval we have developed transmitters and receivers which can bridge the interplanetary gulfs. Even now, if it was felt worthwhile, we could build a transmitter that could send signals to the nearest stars.

Interstellar communication should, therefore, be child's play to a really advanced technological culture. For ages, in all probability, messages of greeting have descended upon the deaf and heedless Earth. In a few years—or a few centuries—we may learn to read them.

It is only within the last five years that we have discovered a better means than radio of speaking across space, in the form of that revolutionary device, the laser, which produces practically parallel beams of light, thousands of times more intense than the Sun. Even more startling developments must, surely, still lie ahead for our infant race to discover.

By one means or other, perhaps a hundred years from now, we may therefore hope to establish contact with other beings. The probability is that they will be much higher in the scale of creation than we are—and this means not only scientifically but spiritually. (As our own species is in the process of proving, one cannot have superior science and inferior morals. The combination is unstable and self-destroying.)

In the past, almost all intellectual progress has arisen from contact between different races and cultures. No nation is sufficient unto itself; there are plenty of examples to show that isolation means, eventually, stagnation. Within a fleeting moment of historic time, there will be but a single culture on Earth, with all men linked together by global communications networks so efficient that space on this planet is effectively annihilated. We shall be in need, then, of fresh philosophies and insight; we may receive them from minds which have been brooding upon the problems of the universe since the great reptiles roamed our world.

It is a strange thought that purely scientific technologies will eventually put us into direct contact with beings with most of the attributes our religions have given to the gods. The contact will be overwhelming; it may be devastating, inducing a kind of inferiority complex that may lead to loss of the will to live—as has happened many times on this planet in the past. Perhaps our eventual fate may resemble that of intelligent gorillas, doomed to look out with dimly comprehending eyes from the bars of our planetary zoo, at a universe largely beyond our understanding.

This is a risk which we have to face; to turn back at this stage would be impossible, and a negation of all the attributes that have made us human. The comforts of ignorance are ephemeral; we shall insist on learning the truth—however unflattering it may be—about our place in the hierarchy of cosmic intelligence, in the spectrum of the spirit.

All of which leads to a most ironic conclusion. At this moment, the drive into space is being conducted by the two most powerful nations on Earth, for reasons which are largely materialistic—prestige, defense, the development of new industries. But in the long run (and perhaps the short run) these things will be utterly unimportant; for the illusions of our day cannot survive the fierce, hard light that beats down from the stars.

Though men and nations may set out on the road to space with thoughts of glory or of power, it matters not whether they achieve these ends. For on that quest, whatever else they lose or gain, they will surely find their souls.

Class of '00

Toward the end of the Middle Ages, so it is said, the better colleges of England could teach addition and even subtraction, but *not* multiplication and division. Advanced students who wished to learn these esoteric arts had to go to the Continent.

As an Englishman, I hope that this story is slightly exaggerated, but it demonstrates the point I wish to make. There are, from time to time, fundamental advances in methods of thinking, and hence of teaching. The invention of decimal notation was one; today, any junior arithmetic class can perform feats that would have seemed miraculous to the ancients.

What future breakthroughs may we expect? There is no doubt that we need them badly. On the whole, the art of the teacher has changed remarkably little since Plato founded the Academy. Yet now we live in a world of inconceivably greater complexity—one in which the sum total of knowledge doubles before a child can grow to maturity. In an age of exploding information, we cannot hope to survive unless we vastly improve the efficiency of our teaching methods.

Teaching is a form of communication, and any advance in communication techniques is quickly reflected in education. (Consider the impact on teaching of the alphabet and the printing press!) Today, with such satellites as Telstar and Syncom, we see the primitive beginnings of a truly global communications network. It will not be many decades before all the present barriers (economic and technical) to planet-wide radio and TV are down, and the windows of every classroom open on the entire world.

Although movies and film strips are already valuable aids to education, the directness and *immediacy* of TV will make an even greater appeal. Think, also, of the effect on the teaching of languages, when classes in different countries can talk to each other across the face of the planet—and can read, projected on their screens, the current newspaper from any land they choose.

Looking still further ahead, we can imagine the time when classes may be literally dispersed over the whole earth, so that the best teachers and the best pupils will no longer be sundered by accidents of geography. Most of these classes will be like today's TV educational programs—that is, one-way only, with a lecturer addressing an unknown and invisible audience. However, for special purposes two-way communication will be possible, as the "Radio Schools" of the Australian Outback have demonstrated for many years. Here, one teacher may have twenty pupils in a ten-thousand-square-mile area—each with his own receiver/transmitter. There may be few cases where such an arrangement would be economical with TV, and on the global scale, but the cases that *do* arise will be extremely important ones. There has already been one transatlantic medical seminar *via* Telstar, which hints at the shape of things to come.

Communications satellites will also help to make possible electronic library and information systems, whereby any document anywhere can be reproduced on demand by a facsimile process. At first, such services will be used largely by specialists and researchers, but someday every high school boy will be able to dial his way, without too much human help, through the memory banks of the British Museum and the New York Public Library—and the yet greater electronic "World Brain" into which these will ultimately be integrated.

It is now obvious that computer-type devices of many kinds will soon invade the classroom, with the result that teachers will be relieved of their duller and less productive chores, such as stuffing raw facts into the minds of their pupils and marking examination papers. Today's "teaching machines," crude though they are, point unmistakably to the future. The time will come when they too will have access to all libraries and information services, and will be able to converse intelligently with their questioners. Then the human educators can concentrate on their real job—that of arousing the imaginations of the young, and inspiring them to seek worthwhile goals.

Yet all these new devices, wonderful and valuable though they may be, do not really get to the heart of the problem. True, they will help the teacher by extending his resources and improving his efficiency. But the mind of the pupil—even the most apt and willing pupil—can absorb information only at a limited rate; the bottleneck of the human sensory channels remains unbroken.

Let us see what this bottleneck really means. When we are listening, we can absorb roughly a hundred words a minute. Our eyes can do perhaps four times as well as this with ordinary text. (Erle Stanley Gardner, say, not Emmanuel Kant.) The invention of writing, therefore, *greatly increased the rate at which we can acquire information*. It is claimed that rapid reading courses can produce further improvements, but hardly of an order of magnitude. We cannot even read one book per eye, which would seem a modest enough ambition. The switchboard of the brain becomes too quickly saturated and will accept no further calls.

Can we ever do anything about this, perhaps by impressing information directly on the brain at high speed? Let me give an analogy from sound recording. If you have a one-hour symphony on tape, you can make a copy of this by re-recording it in "real time"—that is, by taking exactly one hour to play it from one recorder to another. However, you can also do the job in a few minutes, by running both machines at high speed. Provided that certain technical requirements are met, there is no way of distinguishing between the two final products.

This analogy is very suggestive, but it is false in one important detail. Though we have learned a great deal about the human brain in the last few decades, we do not yet know how it stores information. However, we are fairly sure that it does *not* store it like magnetic tape, with each "bit" of information in one unique spot. Men have lost large areas of their brains with no apparent effect on memory; but if you snip a piece anywhere from a magnetic tape, something is irretrievably lost.

The brain is thus more than the sum of its parts, but one day we may discover how those parts work together, and how the interlocking cells of the cortex store our memories. After all, during the past ten years we have discovered how *we* are stored at conception, by the coding of four simple groups of atoms along the double spiral of the DNA molecule. This discovery is quite as astonishing, and quite as fraught with awesome consequences, as would be the unraveling of the brain's recording system.

There seems nothing unreasonable, therefore, in the idea that we may one day be able to implant information directly in the brain, perhaps by some electro-hypnotic technique which would bypass the narrow gateways of the senses. But would such information be really "knowledge," available on demand even though one might never remember learning it in the usual way? That is an intriguing philosophical question, and leads us straight to the problem of memory retrieval—or, if you prefer the term, "mental playback."

Most people are now familiar with Dr. Wilder Penfield's famous discovery that under electrical stimulation of the cerebral cortex, subjects can remember, in fantastic detail, events that happened years ago. There is some evidence that the brain never forgets anything, and that its potential capacity is virtually unlimited. Unfortunately, most of us have very bad filing systems; we have the whole of our past around somewhere, but can't just lay our hands on it at the moment.

Perhaps one day we will be able to do so, using techniques which may be partly mechanical, partly mental; when this happens, the old phrase "To put on one's thinking cap" may be literally true. Our descendants may be able to roam the infinite corridors of their minds as through a well-indexed library, locating, if they wish, every impression and every fact that has ever come to their attention, however fleetingly.

If this seems fantastic, consider the quite unbelievable powers which we already know our minds to possess. A dramatic example of this is given by calculating prodigies —many of whom, contrary to general belief, have been very able and intelligent men. Gauss is a classic case; a contemporary example is Dr. Aitken, professor of mathematics at the University of Edinburgh. He was recently asked to divide four by forty-seven, and reeled off some thirty places in a few seconds.

In our age of electronic computers, this particular skill is of little value, but if the mind can do such things, what other equally amazing abilities are concealed in the brains of every one of us, waiting to be called forth?

The long-term future of education may well lie in

discovering and exploiting these powers. Associated with them, perhaps, are the so-called paranormal abilities, like telepathy, which everyone will doubtless start to take more seriously now that the Russians have begun to do so. (It has just been revealed that at least two groups in the U.S.S.R., with extensive experimental facilities, are working in this field.) There may be nothing at all in ESP; but, on the other hand, it may ultimately open up such gateways into the mind that all our existing methods of imparting knowledge become obsolete. The apotheosis of education would be direct and total rapport between the mind of the teacher and the mind of the pupil. Even with our present primitive techniques, something like this is occasionally achieved for fleeting moments.

These are the moments which make a teacher's life worthwhile, and he should welcome all the aids—electronic, psychological or otherwise—which may bring them more often. Certainly he should not resent the gifts that, in the years ahead, science will offer to education. He should scrutinize them with care; but he should also recognize when the time has come to stop dividing MCCCLVXXVI by LXVIII.

The Meddlers

From his simian ancestors, man has inherited an insatiable itch to meddle with his surroundings. There is a straight and unbroken line of evolution between a cageful of monkeys in the zoo, and the Atomic Energy Commission in the Pacific.

Now a certain amount of meddling is an excellent thing; it laid the foundations of experimental science and of modern technology. But the intelligent meddler must abide by a few common-sense rules, of which the most important are:

1. Do not attempt the unforeseeable.
2. Do not commit the irrevocable.

Though these rules have often been broken, in the past it seldom mattered; for the damage was confined to the meddler and his immediate vicinity. This is no longer the case; the consequences of meddling are now global, and will soon be astronomical.

I have no wish for my typewriter to add to the literary fallout on Fallout, but my first example has to be the Bravo explosion of March 1, 1954, which showered radioactive coral upon the trawler *Lucky Dragon*—miles outside the "safety zone" confidently established by the meteorologists. In many ways, this event set the pattern for the future; those responsible were embarrassed, and hurried to compensate the injured, but showed no particular signs of remorse. Too bad about those fishermen, but little sacrifices like that have to be made for the safety of the United States . . .

Then followed the long dialogue of hypocritical self-interest between the U.S.S.R. and the U.S. on the subject of bomb testing, each claiming the right to contaminate the Earth in pursuance of its policy of massive suicide. As a result, every living human being is now appreciably more radioactive than his grandparents—with incalculable effects upon all the generations to come. Contrary to the science-fiction writers, fallout will not produce a crop of monstrous mutants; extreme variations from the norm have little chance of survival, and less of reproduction. But it will produce an endless series of minor defects, illnesses, and premature deaths which, all told, will add up to a staggering sum of human misery.

Two centuries ago Nathan Hale might regret that he had but one life to give for his country; today's patriots must ask themselves how many genes (and whose) they are prepared to give for theirs. And although the U.S.–U.S.S.R. pact on bomb testing is a welcome step toward sanity, who can say how much damage has already been done?

Quite apart from fission products, our modern world is drenched with chemicals which did not exist ten or twenty years ago. Almost all of them—DDT and the other insecticides, penicillin and its related "wonder drugs"—involve some degree of risk. In most cases, we accept these risks willingly; penicillin has saved thousands of lives for every one jeopardized by allergic reactions—pets and people may have been poisoned by DDT, but it has eliminated typhus and malaria from whole countries. No one but a madman would deny these benefits, yet we must never become complacent and overconfident. Rachel Carson's strident warning, in *Silent Spring,* was necessary, even if exaggerated, though E. B. White saw the danger years before, in his unforgettable *The Morning of the Day They Did It.* That satirical fantasy, now rapidly coming true, described a world where the chemists had made agricultural products so plentiful—and so toxic—that everyone had to take regular injections to counteract the lethal effects of food.

The terrible Thalidomide disaster has alerted everyone to these dangers, for the moment. It has been pointed out that if Thalidomide had been developed in the United States, instead of Europe, "the marketing techniques of the pharmaceutical industry, which can saturate the country with a new drug almost as soon as it leaves the laboratory, would have enabled Thalidomide to produce thousands of deformed infants." (Helen B. Taussig, August 1962 *Scientific American*). The United States escaped this catastrophe by good luck and timely warning of Dr. Kelsey; next time, it may not be so fortunate.

For there will be a next time—though no one knows where and when. The price of safety, as of liberty, is eternal vigilance. The people to watch are the pharmaceutical firms out for a quick buck, and the defense scientists out for a big bang.

Not that nuclear explosions are the only global nuisances committed, or attempted, in the name of security. Perhaps you never heard about Project West Ford, the bright idea of M.I.T.'s Lincoln Laboratories to put a third of a billion tiny radio antennas into orbit. When they learned about it, the world's astronomers reacted with near-unanimous violence, protesting that this cloud of minute satellites would interfere with many types of fundamental research for an indefinite period to come. Despite an appeal by the International Astronomical Union to the United States Government, the experiment went ahead in October 1961.

The first attempt failed, but success was achieved in May 1963. There are rumors of other launchings; a recent issue of the authoritative space journal *Astronautics* comments on an unexplained United States Air Force satellite with these ominous words: "It is difficult to avoid the conclusion that the Air Force is quietly placing additional dipoles in orbit."[1]

The most controversial, and widely criticized, of all space experiments took place in mid-Pacific on July 8, 1963, when—despite a series of launching mishaps that would have discouraged less devoted experimenters, the AEC and the Department of Defense detonated a megaton bomb two hundred miles above Johnston Island. (Sociological note: in the press releases, it's always a "nuclear *device.*" I say it's a bomb, and I say the hell with it.) Once again, there had been a chorus of protests from scientists all over the world; once again, the objectors were made to appear alarmists by bland official statements. There was not the slightest risk, everyone was assured, that the Van Allen belts, which

---

[1] It is only fair to report that this story has a happy ending. Before the second West Ford (originally known as "Needles") experiment was carried out, it was fully analyzed by international committees of leading scientists. They predicted the conditions under which it would be harmless, and the authorities responsible conducted it accordingly. Let us hope that this sets a precedent for future experiments of a global (or cosmic) nature.

had been around for several billion years, would be blown up within five years of their discovery.

Well, the belts are still there, though somewhat groggy. The confident calculations were out by a factor of ten, possibly a hundred. (The argument is still in progress.) Three artificial satellites, placed in orbit at enormous expense, were promptly silenced—or at least muffled—by the unexpectedly powerful blast of radiation. One of them happened to be the very first British-built satellite, kindly launched only a few weeks earlier by the United States Space Administration as part of its well-intended program of international cooperation.

I can only mention in passing (and passing is what we probably are) such "Coming Attractions" as the Neutron Bomb, laser heat rays, and the *really* virulent diseases that the biological warriors will be able to design, when the genetic code has been cracked and we can create organisms that nature never imagined. One would expect such activities to cause trouble; but unfortunately, even "Harmless" experiments, on the scale at which we are now operating, may lead to most peculiar and obscure disasters. For example: The only thing that protects you from a painful death by acute sunburn is a thin layer of ozone, twenty miles above your head. The amount involved is very small, but it almost completely absorbs the sun's lethal ultraviolet rays. Now, in the course of our space experiments, we are dumping enormous quantities of exotic chemicals into the upper atmosphere—quantities which, in some cases, will exceed the amounts of gas already there. This is contamination with a vengeance, and no one knows what its results will be. A generation from now, that ultraviolet may start leaking through the ozonosphere roof, and we'll have to move underground.

Where is this going to lead, as our powers over nature, but not over ourselves, continue to increase? If we extrapolate the present trends in technological megalomania, arrogant ignorance, and national selfishness, this is the type of press release we may expect from the Pentagon, around about the year 1990:

As there has been much ill-informed criticism of the U.S. Space Force's proposed attempt to extinguish the Sun by means of the so-called "Blackout Bomb" (Operation Pluto), the following statement is being issued to reassure the public.

The experiment is based on the discovery by Spitzer, Richardson, Chandrasekhar and others that the injection of polarized neutrinos into a certain class of sunspot can start a chain reaction, which will cause a temporary quenching or damping of the solar thermonuclear process. As a result, the Sun's brilliance will rapidly decrease to about a millionth of its normal value, then recover in a period of approximately thirty minutes.

This important discovery has grave defense implications, for a potential enemy could utilize it to make a surprise attack on the United States under cover of artificially induced darkness. It is obvious, therefore, that for its own security the U.S. must investigate this phenomenon, and this can be done only by a full-scale experiment.

Though it is appreciated that Operation Pluto will cause temporary inconveniences to large numbers of people—a fact deeply regretted by the U.S. Government—the defense of the Free Solar System permits of no alternative. Moreover, the benefits to science will be enormous, and will far outweigh any slight risks involved.

The numerous protests raised against the operation by many foreign scientists are ill-founded, being largely based upon inadequate information. In particular, the attacks launched by Lord Lovell of Jodrell and Sir Fred Hoyle appear to be inspired by political rather than scientific motivations. It is felt that their views would be altogether different if the United Kingdom possessed vehicles capable of carrying suitable payloads to the Sun.

As these critics have suggested that the Sun's recovery time may be of the order of years rather than minutes, a full study of the blackout process has been carried out by the Los Alamos **PHOBIAC** computer. This has shown that the risk of the Sun remaining extinguished is negligibly small, though the actual figure must remain classified.

Nevertheless, to explore all possibilities, the U.S. Government has commissioned the well-known firm of independent consultants, Kahn, Teller, and Strauss to make a study of the situation should the Sun fail to return to normal. Their report—to be released shortly under the title *Economic and Other Effects of a Twenty-Four-Hour Night*—indicates that, though there may be a difficult transition period, the community will soon adapt itself to the new conditions. These may, in fact, be advantageous in many respects; for example, the enormous stimulus to the electrical supply and illumination industry would remove any danger of a recession for years to come.

The protracted absence of the Sun would also render useless the Soviet Union's announced intention of increasing agricultural production, by tilting the Earth's axis so as to move Siberia into the tropics—a proposal which has rightly aroused the disapproval of the civilized world. Should Operation Pluto have unexpected aftereffects, there will, of course, be no tropics.

The United States Government, however, is confident that no such mishaps will occur, and is proceeding with the operation in full consciousness of its global responsibilities. It will not be deflected from its plain duty either by uninformed criticism, or such temporary setbacks as the recent destruction of the planet Mercury by the premature detonation of the first blackout device. This accident has been traced to a piece of chewing gum in the inertial guidance system, and all necessary steps have been taken to prevent its recurrence.

Farfetched? I'm not so sure. For a long time, many of us have been wondering why certain types of stars occasionally blow up; and just recently, astronomers discovered an exploding *galaxy*. By the standards of the universe, our meddling may still be pretty small-scale stuff.

But we're certainly working hard at it; and the best, I'm afraid, is yet to be.

The Lunatic Fringe

The lunatic fringe has always been with us. In every age, there have been people who were willing to believe anything so long as it was sufficiently improbable. Religion, economics, science, politics have all had—and still have—their financial minorities who devote their fortunes, their energies, and often their lives to the cause they have made their own.

Often the cause is a sensible one but its advocates are not; they show that humorless monomania, that inability to see any other point of view, that distinguishes the crank from the enthusiast. One does not have to look very far for examples; the best publicized (perhaps over-publicized) specimens in the United States at the moment are probably the John Birch Society and the Black Muslims.

The driving force behind all such extremist groups and crackpot organizations is a mixture of fear and ignorance. It may be, as in the above cases, a well-justified fear of the Communists or the Ku Klux Klan, but often it stems from deeper and less rational causes. We can see this very clearly if we look at two of the most famous examples of mass moronity in the past decade—Bridey Murphy and the flying saucers. (If Hollywood fancies this title, it can have it.)

In the Bridey Murphy case, a Colorado housewife "remembered" her life as a girl in Ireland more than a century before, and gave an elaborate account of it while under hypnosis. This was built up, by skillful publicity, as evidence of reincarnation, despite the fact that a little careful research would have revealed the truth. When a few skeptical newspapermen did this research, and uncovered the childhood sources from which the subject obtained her memories, the whole sensation collapsed almost overnight.

Let me make one point quite clear. The Bridey Murphy affair involved a perfectly genuine and still unexplained phenomenon almost as remarkable, in its way, as true reincarnation. But this phenomenon—hyperamnesia, or the incredibly detailed and *creative* recall of long-forgotten memories under hypnosis—had been familiar to all psychologists since at least the time of Freud. To have placed it before the American public as proof of survival after death was an act of irresponsible incompetence; some would use stronger terms.

Yet the public lapped it up; the book became, not merely *a* best seller, but *number one best seller,* and several hundred thousand copies were soon in circulation. The publishing trade can take little pride in such exploitation of fear and ignorance—in this case, fear of death, and ignorance of psychology.

The flying saucer craze lasted much longer and indeed is still with us, so there is no need to go into details. But once again, as with Bridey Murphy, we must distinguish between a real phenomenon and the conclusions drawn from it by anxious and hysterical people.

"Unidentified flying objects," to give them their noncommittal name, are quite common; if you have never seen a UFO, you should be ashamed of yourself, for it means that you are not very observant. (I've encountered seven, including two that would have convinced the most skeptical.) They have dozens of causes, many of them ludicrously simple, for it is amazing what nature can contrive when she is in the mood—look at the rainbow or the snowflake. A small proportion of UFO's has never been satisfactorily explained, and the theory that they are visitors from outer space is a perfectly reasonable one; I would be the last to condemn it, since I have spent most of my life expounding the possibility.

What I *am* condemning is the credulous naïveté of those who have accepted this theory and made almost a religion of it. On the strength of a few faked photographs and the ravings of obviously psychopathic personalities, thousands were convinced that men from space had actually landed on this Earth. Many still believe this, despite the fact that ever since the opening of the International Geophysical Year the skies of our planet have been raked by armies of trained observers and every conceivable type of detecting instrument.

The chance of a genuine spaceship evading discovery in this age of multibillion dollar radar networks and Moon-watch teams is about the same as that of a dinosaur concealing itself in Manhattan.[1] When our stellar neighbors really do start to arrive, we'll know all about it within five minutes. The idea that any government could—or would—keep such a world-shattering event secret *for year after year* is utterly ludicrous.

The fears of the UFO-ologists are more complex than those of the Bridey Murphy believers. (I would guess, by the way, that the two groups overlapped to a very large extent, for credulity knows no boundaries.) Alarm at the drift to atomic destruction was combined with the hope that benevolent saviors from the sky would arrive and tidy up the mess we have made of this planet. And in this case, the ignorance which made so many honest people misinterpret the evidence of their own eyes was completely excusable. Not even the scientists had realized what an extraordinary collection of optical, astronomical, meteorological and elec-

---

[1] I once saw a movie in which this happened. It contained the immortal line: "It's hiding somewhere in the Wall Street area!"

trical apparitions inhabited our skies. The UFOs have done some good by focusing attention upon these.

It is depressing to make a list of the other pseudo-scientific ideas that had achieved fame or notoriety during the last generation. Do you remember Emmanuel Velikovsky's *Worlds in Collision?* This monument of impressive scholarship explained Biblical history on the theory that the Earth and Venus had cannoned together in the past like billiard balls. L. Ron Hubbard's *Dianetics* belonged to the same category; it purported to cure all mental ills by taking the patient back to the moment in time when his trouble started. Another revelation must be imminent, now that the Beats are buried and the zest for Zen is flagging. I have no idea what it will be, and am in no great hurry to find out.

You may feel that this is making too much of something that affects only a small part (one hopes) of the total population. It is true that in the past crankiness and eccentricity did little harm, and even added a certain spice to society. A generation ago, flat-Earthers, end-of-the-World cultists, and disciples of weird religions caused no embarrassment outside their immediate circle. But we are moving now into a complex and perilous age, where credulity and superstition are luxuries that can no longer be afforded. For consider this example:

In 1843, fifty thousand followers of the prophet William Miller gathered on New England hilltops to await the expected hour of judgment. The advent of a great comet, its tail streaming like a fiery banner across the sky, seemed to them a sign that the end of the world was at hand.

Men are still watching the sky for signs of doom; but now they look into radar screens. And here is the important difference; the beliefs of fifty thousand Millerites could have no influence, one way or the other, upon the end of the world, but today, when we can carry the power of Vesuvius in a single warhead, the fears or delusions of only fifty men could bring it about.

This is an extreme case; but all forms of irrationality

are dangerous, because in the right circumstances they can spread like a plague, infecting not only a community but an entire nation. Those concerned may be very ashamed of themselves afterwards, but by then the damage may be done.

You cannot build an informed democracy out of people who'll believe in little green men from Venus. Credulity—willingness to accept unsupported statements without demanding proof—is the greatest ally of the dictator and the demagogue. It is not so very long ago that there were voices crying: "The Jews are plotting against the Reich!" and "I have here in my hand a list of 205 Communists in the State Department." Those voices are silent now; but there will be others.

One of the factors, ironically enough, which has contributed to popular willingness to accept the incredible is the success of modern science. Because so many technical marvels have been achieved, the public believes that the scientist is a magician who can make *anything* happen. It does not know where to draw the line between the possible, the plausible, the improbable, and the frankly absurd. Admittedly this is often extremely difficult, and even the experts sometimes fall flat on their faces. But usually, all that is required is a little common sense.

Unfortunately, common sense has always been rather rare. As a reminder of this, let me quote two final examples of mass stupidity, which may also help to dispel the idea that it is a United States monopoly.

During the darkest days of the First World War, the rumor swept the length and breadth of Britain that troops were arriving from Russia in huge numbers (this was before the Revolution) to reinforce the crumbling western front. Thousands of honest Britons "saw" them at ports and railway stations, and millions believed the rumor, because they wanted to. And how did the observers know that these soldiers were actually Russians? Not because they said so—*but because they had snow on their boots.*

That little detail was the clincher, as far as most

people were concerned. They never stopped to ask if even Russian snow would survive the long sea voyage from Murmansk to Scotland.

My last example may surprise you; you may not know that flying saucers have invaded the Soviet Union. Yet they have, for a remarkable reason. According to *Pravda,* which is rather indignant about the whole affair, Russian saucer fans believe that little people from Venus constantly descend on Uzbekistan and Tajikistan and then "promptly scurry in all directions in search of inexpensive Oriental sweets."

I love that "inexpensive"; presumably, the ruble is hard to get on the Venus black market.

No one should derive much satisfaction from this proof that nuttiness is also rampant on the other side of the Iron Curtain. Unreason is always a menace, wherever it occurs; it may be even more of a danger in the Soviet Union than in a country with democratic safeguards. (Look what Hitler's intuitions did to the world.) And there is, unfortunately, no reliable cure for it; you cannot buy sanity at the drugstore, or inject common sense into the community by mass inoculation.

The only answer lies in education, and even that is merely a palliative, not a panacea, for a college degree is no guarantee of wisdom, as anyone who has ever been near a campus will testify. There are many people in the world who are educated beyond their intelligence, but there are far, far more who have not been educated to within hailing distance of it. They are the ones who provide fodder for the demagogues and cranks, who listen to false prophets and sponsor absurd or evil causes. They cannot always be blamed, for society has robbed them of what should be every man's right—an education to the limit of his ability, whatever his financial status, creed, or color. No wonder that, dimly realizing their deprivation, they seek any substitute that they can find.

Very often that substitute takes the form of anti-intellectualism—a pretense that knowledge, education, and culture are worthless or even dangerous. This is, of course, a typical sour-grapes reaction; not long ago one could identify those suffering from it by their fondness for the word "egghead." That engaging term is now a little out of fashion, because the events of the last few years have made it obvious to everyone that a society which despises brains is on the one-way road to oblivion.

Human nature being what it is, the lunatic fringe can never be abolished—and most of us, if the truth be told, would hate to see it vanish altogether. But education can minimize its influence, can convert it from a potential danger to a source of mild amusement. A century ago, Matthew Arnold compared this world to a "darkling plain . . . where ignorant armies clash by night." The metaphor is still valid. Perhaps the greatest single task that now faces every nation is the dispelling of that ignorance, lest the armies clash again—for the last time.

The Electronic Revolution

The electron is the smallest thing in the universe; it would take thirty thousand million, million, million, million of them to make a single ounce. Yet this utterly invisible, all but weightless object has given us powers over nature of which our ancestors never dreamed. The electron is our most ubiquitous slave; without its aid, our civilization would collapse in a moment, and humanity would revert to scattered bands of starving, isolated savages.

We started to use the electron fifty years before we

discovered it. The first practical application of electricity (which is nothing more than the ordered movement of electrons) began with the introduction of the telegraph in the 1840's. With really astonishing speed, a copper cobweb of wires and cables spread across the face of the world, and the abolition of distance had begun. For over a century we have taken the instantaneous transfer of news completely for granted; it is very hard to believe that when Lincoln was born, communications were little faster than in the days of Julius Caesar.

Although the beginning of "electronics" is usually dated around the 1920's, this represents a myopic view of technology. With the hindsight of historical perspective, we can now see that the telegraph and the telephone are the first two landmarks of the electronic age. After Alexander Graham Bell had sent his voice from one room to another in 1876, society could never be the same again. For the telephone was the first electronic device to enter the home and to affect directly the lives of ordinary men and women, giving them the almost godlike power of projecting their personalities and thoughts from point to point with the speed of lightning.

Until the closing years of the nineteenth century, men used and handled electricity without knowing what it was, but in the 1890's, they began to investigate its fundamental nature, by observing what happened when an electric current was passed through gases at very low pressures. One of the first, and most dramatic, results of this work was the invention of the X-ray tube, which may be regarded as the ancestor of all the millions of vacuum tubes which followed it. A cynic might also argue that it is the only electronic device wholly beneficial to mankind—though when it was invented many terrified spinsters, misunderstanding its powers, denounced poor Röntgen as a violator of privacy.

There is an important lesson to be learned from the X-ray tube. If a scientist of the late Victorian era had been asked "In what way could money best be spent to further the progress of medicine?" he would never by any stretch of the imagination have replied: "By encouraging research on the conduction of electricity through rarefied gases." Yet that is what would have been the right answer, for until the discovery of X rays doctors and surgeons were like blind men, groping in the dark. One can never predict the outcome of fundamental scientific research, or guess what remote and unexpected fields of knowledge it will illuminate.

X rays were discovered in 1895—the electron itself just one year later. It was then realized that an electric current consists of myriads of these submicroscopic particles, each carrying a minute negative charge. When a current flows through a solid conductor such as a piece of copper wire, we may imagine the electrons creeping like grains of sand through the interstices between the (relatively) boulder-sized copper atoms. Any individual electron does not move very far, or very fast, but it jostles its neighbor and so the impulse travels down the line at speeds of thousands of miles a second. Thus when we switch on a light, or send a Morse dash across a transatlantic cable, the response at the other end is virtually instantaneous.

But electrons can also travel *without* wires to guide them, when they shoot across the empty space of a vacuum tube like a hail of machine-gun bullets. Under these conditions, no longer entangled in solid matter, they are very sensitive to the pull and tug of electric fields, and as a result can be used to amplify faint signals. You demonstrate the principle involved every time you hold a hose-pipe in your hand; the slightest movement of your wrist produces a much greater effect at the far end of the jet. Something rather similar happens to the beam of electrons crossing the space in a vacuum tube; they can thus multiply a millionfold the feeble impulses picked up by a radio antenna, or paint a fluorescent picture on the end of a television screen.

Until 1948, electronics was almost synonymous with the vacuum tube. The entire development of radio, talkies, radar, television, long-distance telephony, up to that date depended upon little glass bottles containing intricate structures of wire and mica. By the late 1940's the vacuum tube had shrunk from an object as large as (and sometimes almost as luminous as) an electric light bulb, to a cylinder not much bigger than a man's thumb. Then three scientists at the Bell Telephone Laboratories invented the transistor and we moved from the Paleo-electronic to the Neoelectronic Age.

Though the transistor is so small—its heart is a piece of crystal about the size of a rice grain—it does everything that a radio tube can do. However, it requires only a fraction of the power and space, and is potentially much more reliable. Indeed, it is hard to see how a poorly designed transistor can ever wear out; think of little Vanguard I, still beeping away up there in space, and liable to continue indefinitely until some exasperated astronaut scoops it up with a butterfly net.

The transistor is of such overwhelming importance because it (and its still smaller successors) makes practical hundreds of electronic devices which were previously too bulky, too expensive or too unreliable for everyday use. The pocket radio is a notorious example; whether we like it or not, it points the way invariably to a day when person-to-person communication is universal. Then everyone in the world will have his individual telephone number, perhaps given to him at birth and serving all the other needs of an increasingly complex society (driving license, social security, credit card, permit to have additional children, etc.). You may not know where on Earth your friend Joe Smith may be at any particular moment; but you will be able to dial him instantly—if only you can remember whether his number is 8296765043 or 8296756043.

Obviously, there are both advantages and disadvantages in such a "personalized" communication system; the solitude which we all need at some time in our lives will join the vanished silences of the pre-jet age. Against this, there is no other way in which a really well-informed *and* fast-reacting democratic society can be achieved on the original Greek plan—with direct participation of every citizen in the affairs of the state. The organization of such a society, with feedback in both directions from the humblest citizen to the President of the World, is a fascinating exercise in political planning. As usual, it is an exercise that will not be completed by the time we need the answers.

A really efficient and universal communications system, giving high-quality reception on all bands between all points on the Earth, can be achieved only with the aid of satellites. As they come into general use, providing enormous information-handling capacity on a global basis, today's patterns of business, education, entertainment, international affairs will change out of all recognition. Men will be able to meet face to face (individually, or in groups) without ever leaving their homes, by means of closed circuit television. As a result of this, the enormous amount of commuting and traveling that now takes place from home to office, from ministry to United Nations, from university to conference hall will steadily decrease. There are administrators, scientists and businessmen today who spend about a third of their working lives either traveling or preparing to travel. Much of this is stimulating, but most of it unnecessary and exhausting.

The improvement of communications will also render obsolete the city's historic role as a meeting place for minds and a center of social intercourse. This is just as well anyway, since within another generation most of our cities will be strangled to death by their own traffic.

But though electronics will ultimately separate men from their jobs, so that (thanks to remote manipulation devices) not even a brain surgeon need be within five thousand miles of his patient, it must also be recognized that few of today's jobs will survive long into the electronic age. It is now a cliché that we are entering the Second Industrial Revolution, which involves the mechanization not of energy, but of thought. Like all clichés this is so true that we seldom stop to analyze what it means.

It means nothing less than this: There are no routine,

non-creative activities of the human mind which cannot be carried out by suitably designed machines. The development of computers to supervise industrial processes, commercial transactions and even military operations has demonstrated this beyond doubt. Yet today's computers are morons compared to those that they themselves are now helping to design.

I would not care to predict how many of today's professions will survive a hundred years from now. What happened to the buggywhip makers, the crossing sweepers, the scriveners, the stonebreakers of yesteryear? (I mention the last because I can just remember them, hammering away at the piles of rock in the country lanes of my childhood.) Most of our present occupations will follow these into oblivion, as the transistor inherits the earth.

For as computers become smaller, cheaper and more reliable they will move into every field of human activity. Today they are in the office; tomorrow they will be in the home. Indeed, some very simple-minded computers already do our household chores; the device that programs a washing machine to perform a certain sequence of operations is a specialized mechanical brain. Less specialized ones would be able to carry out almost all the routine operations in a suitably designed house.

Because we have so many more pressing problems on our hands, only the science-fiction writers—those trail-blazers of the future—have given much thought to the social life of the later electronic age. How will our descendants be educated for leisure, when the working week is only a few hours? We have already seen, on a worldwide scale, the cancerous growths resulting from idleness and lack of usable skills. At every street corner in a great city you will find lounging groups of leather-jacketed, general-purpose bioelectric computers of a performance it will take us centuries and trillions of dollars to match. What is their future—and ours?

More than half a century ago H. G. Wells described, in *The Time Machine,* a world of decadent pleasure lovers, bereft of goals and ambitions, sustained by subterranean machines. He set his fantasy eight hundred thousand years in the future, but we may reach a similar state of affairs within a dozen generations. No one who contemplates the rising curve of technology from the Pilgrim fathers to the Apollo Project dare deny that this is not merely possible, but probable.

For most of history, men have been producers; in a very few centuries, they will have to switch to the role of consumers, devoting their energies 100 per cent to absorbing the astronomical output of the automated mines, farms and factories.

Does this *really* matter, since only a tiny fraction of the human race has ever contributed to artistic creation, scientific discovery or philosophical thought, which in the long run are the only significant activities of mankind? Archimedes and Aristotle, one cannot help thinking, would still have left their marks on history even if they had lived in a society based on robots instead of human slaves. In any culture, they would be consumers of goods, but producers of thought.

We should not take too much comfort from this. The electronic computers of today are like the subhuman primates of ten million years ago, who could have given any visiting Martians only the faintest hints of their potentialities, which included the above mentioned Archimedes and Aristotle. Evolution is swifter now; electronic intelligence is only decades, not millions of years, ahead.

And *that*—not transistor radios, automatic homes, global TV—is the ultimate goal of the Electronic Revolution. Whether we like it or not, we are on a road where there is no turning back; and waiting at its end are our successors.

H. G. Wells and Science Fiction

If we define "science fiction" as literature about strange worlds and creatures, departures from the accepted laws of nature, and wonderful inventions which have some scientific plausibility and do not involve sheer magic, then we can trace it back to the beginning of fiction itself. There are science-fictional elements in the Odyssey and the Greek myths, though they are inextricably entangled with pure fantasy.

"S.f." in a recognizably modern form arose, as might be expected, soon after the discoveries of Galileo and Newton. The invention of the telescope, at the beginning of the seventeenth century, was a tremendous stimulus to imagination as well as to science. It hinted at the true scale of the universe, and revealed for the first time that there were other worlds than Earth. Stories of journeys to the Moon started to be popular from about 1630 onward, the most notable examples being Kepler's *Somnium* and Cyrano de Bergerac's *Voyage to the Moon*. (For details of these and similar works, see the opening chapters of Willy Ley's *Rockets, Missiles and Space Travel* or my own *Exploration of Space*.) *Gulliver's Travels*, published in 1726, is pure science fiction; the flying island of Laputa has a very up-to-date ring, and physicists hunting for funds would do well to read Swift's account of the various Laputian research projects.

Throughout the nineteenth century, science-fictional ideas became more and more widespread in the general literature. Famous examples are Poe's *Descent into the Maelstrom,* Mary Wollstonecraft Shelley's *Frankenstein,* Fitz-James O'Brien's *The Diamond Lens,* Robert Louis Stevenson's *Dr. Jekyll and Mr. Hyde,* Rider Haggard's *She,* Lord Lytton's *The Coming Race.* Specialization had not yet set in, and there were no invidious distinctions between "main-stream" and s.f. writers. Indeed, well into this century there were authors who moved freely and easily in both fields; think of Conan Doyle *(The Lost World),* Rudyard Kipling *(With the Night Mail);* Aldous Huxley *(Brave New World)* and E. M. Forster *(The Machine Stops).*

The first writer of world renown to make his name in the field of s.f., and indeed to become identified with it, was Jules Verne. An admirable storyteller, with a keen insight into both the possibilities and dangers of modern technology, Verne anticipated and described many of the inventions which we now take for granted. Even when he did not conceive them himself (there were, for example, submarines already in existence when Verne wrote *Twenty Thousand Leagues Under the Sea),* he understood more clearly than any of his contemporaries that science was about to transform civilization, and that the shape of the future would be very different from that of the past.

The careers of Verne (1828–1905) and Herbert George Wells (1866–1946) overlap; Verne was still alive when Wells published his finest tales. Though there are some parallels between the two men, the differences are much more striking. It cannot be denied that Wells was a far greater writer, owning almost all the gifts that a novelist can possess. Perhaps he had too many gifts; if he had not been so interested in politics, history and society, he might have written fewer but better books. (In fifty years, he produced some 150 titles, of which perhaps twenty are remembered today.)

Wells grew up in poverty and squalor; his father, Joseph, had been a gardener, his mother a lady's maid, but before he was born his parents had sunk their small resources in an unsuccessful shop, which was saved from bankruptcy by Joseph's earnings as a professional cricketer. H. G.'s early years are best summed up in his own words:

> When the writer was ten or eleven his father was disabled by a fall which crippled him, and when he was thirteen the little shop collapsed. His mother returned as a housekeeper to her former mistress and his father took refuge in a small, inexpensive cottage. Further education for the writer seemed impossible. There was some trouble in finding him employment, an unhandy boy preoccupied with reading. He was tried over as a draper's apprentice, as a pupil-teacher in an elementary school, as a chemist's apprentice, and again as a draper. After two years with the second draper he prayed to have his indentures cancelled, and became a sort of pupil-teacher. In the interval between these attempts to begin life he took refuge in the housekeeper's room with his mother.

From this pathetic environment Wells escaped by a combination of luck and genius. He won a scholarship to the Royal College of Science, Kensington, where he studied biology under the great T. H. Huxley and took his degree in zoology. When he was twenty-one, an accident on the football field destroyed one kidney and made him a semi-invalid for a while, with the result that he had both the opportunity and the incentive to write for a living.

He was successful from the very start with short stories, articles and humorous sketches. His first novel, *The Time Machine* (still his masterpiece), appeared in book form in 1895, and thereafter his fame spread swiftly throughout the world. Even the miseries of his early life were turned to good account in such novels as *Kipps, Tono-Bungay* and *The History of Mr. Polly*. These sagas of ordinary people in turn-of-the-century England belong to the best tradition of Dickens, and would have assured Wells's fame even if he had never trafficked with Martians.

In the preface to the 1924 Atlantic Edition of his works, H. G. Wells wrote: *"The Invisible Man* was first published in 1897, and *The War of the Worlds* in 1898. There is very little to be said about either work. They tell their own stories."

The last sentence is true enough; both novels are splendid adventure tales which can be read for the sheer enjoyment of their fast-moving action. But, Wells to the contrary, there is a great deal that can be said about them—much more than the author could ever have imagined.

The Invisible Man is the slighter work, and is not science fiction but sheer fantasy, despite Wells's ingenious attempt to make the basic idea[1] plausible by talking about refractive indices. The interest of this story lies, however, not in its scientific concepts but in the brilliantly worked out development of the theme of invisibility. If one *could* be invisible—what then?

At first sight, if one may use so unsuitable a phrase, an invisible man would seem to be omnipotent. He could go everywhere and do anything, and the world would be helpless to oppose his plans. But even at the opening of his literary career, Wells knew better than this, and shows that an invisible man should be pitied as much as feared. His great gift would really be an intolerable curse, from which he would do his utmost to escape.

Wells did a variation on this theme in the most famous of his short stories, "The Country of the Blind," using as his opening proverb "In the country of the blind, the one-eyed man is king." He demolished this statement as thoroughly as he disposed of the invisible man's invincibility; it is fascinating to compare the two tales and to study their approaches to the same central idea.

---

[1] I have discussed this in some detail in the chapter "Invisible Men and Other Prodigies" in *Profiles of the Future, An Inquiry into the Limits of the Possible.*

The War of the Worlds is in some ways Wells's most remarkable tour de force, and contains passages whose relevance is far greater today than when it was written at the close of the last century. The original idea was not due to Wells but to his older brother Frank:

> a practical philosopher with a disbelief even profounder than that of the writer in the present ability of our race to meet a great crisis either bravely or intelligently. . . . Our present civilization, it seems, is quite capable of falling to pieces without any aid from the Martians. [Preface to Atlantic Edition.]

This astonishing novel contains what must be the first detailed description of mechanized warfare and its impact upon an urban society. Yet Wells wrote it not only before the First World War, but even prior to the Boer War! The account of refugees streaming out of London before the assault of the Martians must have seemed unbelievable fantasy to the comfortable Victorians; to us, it is more like straight news reporting. We have seen it happen, and we know that it can happen again.

Every generation can re-read *The War of the Worlds* in the light of its own experience, and gain something new from it. In the 1920's it was impressive because it described poison gas in action, and suggested that aircraft could be used for warfare. It made a still greater impact on the 1930's, when the famous Orson Welles's production of Howard Koch's radio script (Mercury Theater of the Air, C.B.S., October 30, 1938) caused panic over a substantial area of the eastern United States. Exactly the same thing happened a few years later in South America, and I have no doubt that the experiment could be successfully repeated today—such is the power of the myth that Wells created.

As we have seen, there had been many earlier tales of interplanetary travel, and even a few of extraterrestrial visitors to Earth. These cosmic tourists were usually friendly or superior beings whom the author used as commentators on human affairs; the Menace from Space (capitals seem unavoidable) was unknown before the time of Wells. It has, alas, been all too common since.

However, unlike his later imitators, Wells made his monsters plausible and well-motivated. They were not hell bent on destruction for its own sake, but were proceeding on a logical plan of conquest, with a definite though deplorable—from our biased point of view—objective. A good example of Wells's foresight is the hint, forty years before Quisling and Co., that there would be men quite willing to work out an accommodation with the Martians.

Though one cannot blame Wells for all the later excesses of interplanetary warfare, perhaps he merits some criticism for propagating the creed that anything alien is likely to be horrible. Compare this unflattering description of the Martians ("Those who have never seen a living Martian can scarcely imagine the strange horror of its appearance. . . . Even at this first encounter, I was overcome with disgust and dread. . . .") with the passage in C. S. Lewis's *Perelandra* where the hero meets a much more imposing monster in a Venusian cave—and, after the first revulsion, sees it merely as something strange, not hideous at all. The underwater explorers of today have been through the same process while making friends with octopi, and the astronauts of the future may have similar problems. The tradition started by *The War of the Worlds* will not help them; even while writing this preface, I came across the suggestion by an American general that the first lunar expedition should be armed. He may be quite right; and so may Wells—in which case, heaven help us all.

For the reader in the 1960's, *The War of the Worlds* contains two other items of special significance. The description of the remote-controlled Martian handling machines bears an altogether uncanny resemblance to the mobile robots now being developed for operations in dangerous or inaccessible environments. And secondly, there is the heat ray.

If Wells did not invent this terrifying weapon, he certainly launched it upon its career of fictional destruction. Unlike the atomic bomb, which he used and named in his 1914 novel *The World Set Free*, the heat ray has remained on paper, and until recently seemed likely to stay there. We cannot count on this much longer; once again, reality is about to catch up with Wells. The scientific breakthrough which made death rays possible took place around 1960; if you have not yet heard of infrared lasers, I am afraid you will soon do so.[2]

For the 1970's, if there are any 1970's, *The War of the Worlds* will carry the gravest message of all. Wells was almost certainly the first man to foresee the possibility of extraterrestrial contamination, which is already a matter of alarm to space scientists. At the moment they are worried that the Earth will infect other planets; in the 1970's they will be still more concerned with traffic in the other direction.

After all this, those who have not read *The War of the Worlds* may be surprised to know that, like much of Wells's writing, it is full of poetry and contains passages that catch at the throat. Wells tried to pretend that he was not an artist, and stated that "There will come a time for every work of art when it will have served its purpose and be bereft of its last rag of significance." This has not yet happened for the best of Wells's science fiction, though it has done so for all but a few of his realistic and political novels. These have suffered the fate of most "topical" writing, while his so-called fantastic tales are still fresh and enjoyable.

There may, of course, be some readers who cannot accept the pre-1900 setting of the stories in this volume, and feel that Martian invaders are incompatible with hansom cabs. It is true that George Pal and Howard Koch quite legitimately updated *The War of the Worlds* for the movie and radio versions, and also transferred it to the United States. (The Martians were remarkably

---

[2] This prediction has already been amply fulfilled. See, e.g., *Goldfinger*.

good shots to have landed all their projectiles around London; just how good, we are now in a position to appreciate as our own space probes go wandering off to odd points of the solar system.) But the novel must remain as Wells wrote it, frozen in time and space, and giving a glimpse of a world now almost as strange to us as the Martians were to Victorians. Only those of little imagination, who have no right to be reading this book in the first place, will be worried by the nineteenth century setting.

A year later, in 1899, Wells wrote his second interplanetary romance, *The First Men in the Moon*. This is perhaps the most famous of all stories of space travel; only three later writers ever came near to matching its mood of extraterrestrial awe and wonder, and no one has surpassed it. *The Sleeper Wakes, The War in the Air, The Food of the Gods* followed in quick succession, interspersed with dozens of short stories in which Wells mapped out territory since explored by two generations of science-fiction writers. Then, for more than twenty years, Wells virtually abandoned the genre that had brought him fame, though he returned to it toward the end of his life in *Star-Begotten* (1937). His last book, *Mind at the End of Its Tether*, was a sad and despairing work published only two years before his death at the age of eighty. It did his reputation little good, and during the postwar period there was a definite slump in Wells's stock.

Now, however, this mood has passed. A flourishing H. G. Wells Society has been formed in England, and critical books on Wells are appearing in increasing numbers (e.g. Bernard Bergonzi's *The Early H. G. Wells*, W. Warren Wagar's *H. G. Wells and the World State*). This may be partly due to the shamefaced realization in

---

[3] These three, on widely differing literary levels, are C. S. Lewis, Stanley G. Weinbaum, and Edgar Rice Burroughs. Only s.f. fans will remember Weinbaum, who died young, a year after the appearance of his first story. As for the much underrated Burroughs (no man who created the most famous character in modern fiction can be altogether ignored) you may have to take my word for it. If you are over seventeen, it is already too late to read him.

literary circles that Wells's scientific romances were not youthful aberrations or escapist fantasies, but works of art with unique relevance for our times.

And I think that people are re-reading Wells because they are tired of ever-more-minute dissections of neurotic egos, and worn-out repetitions of eternal triangles and tetrahedra. Wells saw as clearly as anyone into the secret places of the heart, but he also saw the universe, with all its infinite promise and peril. He believed—though not blindly—that men were capable of improvement and might one day build sane and peaceful societies on all the worlds that lay within their reach. We need this faith now, as never before in the history of our species.

"Dear Sir . . ."

One of the occupational hazards of authorship, not usually regarded as a high risk profession, is the Letter from the Reader. I think that I can speak with fair authority on this subject, having seriously jeopardized my amateur status by publishing, at last count, thirty-four—oops!—thirty-five books, as well as about four hundred articles and short stories. The result of this garrulity is a fine collection of foreign stamps and a thick file of letters from every part of the world, including the South Pole. Some of these letters have not been easy to answer, or even, for that matter, to read.

And I *do* answer them, for I feel that anyone who troubles to write to an author deserves the courtesy of a reply. However, it is a brief one, for I have long been haunted by the fate of the late H. P. Lovecraft. In case you have never heard of him, Lovecraft was a talented fantasy writer of the 1920's, who slowly starved to death while conducting a gigantic correspondence of thirty-page letters with about a hundred friends and acquaintances. He probably would have starved to death anyway on half a cent a word; but I have no intention of repeating his tragic error, and it is rare for my replies to extend beyond one paragraph.

In most cases, no more is necessary. Requests for autographs, corrections of errors (invariably, of course, the fault of my secretary or the printer), gratuitous information—a bare acknowledgment is sufficient for these. A little more thought is required to deal with one recent—as far as I'm concerned—phenomenon: the class assignment. I occasionally get a batch of letters beginning: "Miss Jones, our English teacher, has asked us to write an essay about science fiction, and has suggested that you. . . ." Good for Miss Jones; the only snag is that when these painfully composed epistles reach me by sea mail, at least six weeks have elapsed, and back at P.S. 473 my name is mud. So I always begin with a careful explanation of the delay, in hopes that the damage can still be undone. Every author should be extremely kind to teenage readers; they are insurance for his declining years.

Where youngsters are concerned, it is particularly difficult to deal with letters—nay, parcels—containing elaborate plans of space rockets, together with endless pages of explanation. I treat these gently, having been through the same phase myself around the age of fourteen. It is fairly obvious that a boy cannot compete with a design team of ten thousand scientists and half a billion dollars' worth of computers, which is roughly what it takes to produce the paper work for a large space vehicle, and most of these hopeful plans can be dismissed as sheer nonsense at a glance. (The entire fuel tankage, for example, may be tucked under the pilot's seat where it won't be in the way.) But I should hate to discourage any future von Brauns, and all these efforts show enthusiasm, application, and the strength of mind to ignore the TV screen for several hours on end. My reply is, therefore, noncommittal as far as the specific design is concerned, but spends some time emphasizing the training and experience needed before one can do anything useful in the space field. It ends with a short list of reference books and magazines, and a few words of encouragement.

No encouragement at all is received by helpful characters who can think of brilliant stories, but just haven't time to write them down. They are prepared to hand their brain children over to me, for 50 per cent of the take.

I have yet to receive a single really worthwhile idea or plot in this way. When the concept is good, it has invariably been used before, and I pride myself on being able to say at once: "I'm sorry, Mr. Smaltz, but Sam Fink published a story about giant man-eating hamsters in *Flabbergasting Fiction* for May 1932."

In any case, being an independent sort of guy, I hate to use someone else's ideas, even when they are both good and original. Twice in my life I have used plots donated by personal friends, and that was years ago. Gifts from strangers should be regarded with particular suspicion, for any author who accepts them may be handling stolen property. There is no certain defense against accidental plagiarism, but why increase the risk?

So far—touch wood—I have never been accused of this ultimate literary crime, but I have had one near miss and just a few weeks ago I was shaken when another author, Poul Anderson, brought out a story on a theme which I fondly believed I was the first to develop. Much more remarkable, we had both concocted the same nonexistent word—*Sunjammer*—for the title. This in itself should be enough to disprove any charges of plagiarism (no thief is *that* stupid), but it is just as well that the stories appeared simultaneously.

As for the near-miss, almost ten years ago I made this brief entry in my little black notebook: "Plague of indecent cloud formations." I would have bet any reasonable sum that this idea was unique to my own dirty mind; imagine my utter astonishment in discovering that Philip Wylie had got there first.

These examples have made me all the more determined to refuse pleas from budding authors to read their stories and offer criticism (by which, as Somerset Maugham remarks, they really mean praise). Quite apart from the time and effort that this would involve, I could never be sure that, years in the future, my subconscious might not dredge up some item from an otherwise long-forgotten manuscript. Then the indignant author might rush into print, or court, and would be able to *prove* that I'd stolen the only good idea he ever had.

Perhaps the most unusual offer I ever received from a reader was the loan of his name. He would be very happy, he said, to appear as a character in one of my stories. I accepted the gift with enthusiasm, explaining that I'd been completely stuck over the name of an interplanetary beatnik who made a living smuggling narcotic baby food from Mars to Earth. However, as the writer did not return the legal clearance forms I sent him, the project got no further.

One mail-borne menace against which I have never developed an adequate defense is that famous American invention, the questionnaire. It is particularly insidious when it asks interesting questions, because then I feel a compulsion to answer them. ("What writer has had the greatest influence on you?" "Which do you consider your best book?" "What treatment would you recommend for literary critics?" "Can an author overcome the handicap of a normal sex life?" etc. etc.) The trouble with questionnaires is that when a professional writer starts to take himself apart, he may not be able to put the pieces together again. He risks the fate of the centipede in the poem, who was asked to explain his method of locomotion, and thought so hard that he ended up "distracted in the ditch, considering how to walk."

All authors are afflicted by a certain amount of crackpot mail, and considering the sort of stories *I* write, my share of it seems surprisingly low. I have never had a letter from Napoleon; and though I have had two or three from God, there have been none—rather disappointingly—from His opposite number.

At luckily rare intervals there will be a handwritten or horribly typed, thesis on gravity or cosmology, with a letter informing me that the writer has spent ten years developing his revolutionary new theory, which *finally* explains the whole universe. As he can't get "orthodox scientists" to listen to him, he pays me the dubious compliment of assuming that I will.

I never attempt to argue with these people, even when I go to the trouble (which is seldom) of reading their effusions. These are almost always scientifically illiterate, full of fallacies that one could no more explain to their benighted authors than one could teach calculus to a chimpanzee.

The wasted effort involved in these products is sometimes horrifying to contemplate, and I cannot help wondering how these unfortunate peoples' friends and relatives are affected by their activities. This struck me forcibly some years ago, when I received a letter from a gentleman who was convinced that the Earth was shaped like an inner tire, and had a photo of a model that proved it. With the total irrelevance which is so typical of cranks, he also enclosed a photo of his wife and three little daughters. They looked a perfectly delightful family, but I would hate to take any bets on their future happiness. On the other hand, it might well be that father's harmless insanity kept him from a harmful one. Crackpottery may be a useful escape mechanism; I hand this thought over to the psychiatrists—who will doubtless dismiss it as a crackpot idea.

How to deal with letters of this type? Well, I have a series of answers, graded according to circumstances. Once again, they never run to more than one paragraph, even if they are in reply to fifty pages of minute handwriting and an acre of diagrams. They are careful to avoid criticism, and even more careful to avoid the slightest degree of encouragement. A typical specimen is:

Dear Sir,
   Thank you for your letter and enclosures, which I have examined with care. It is obvious that you have spent a great deal of time on this project, and I regret that heavy pressure of work will not allow me to comment upon it at length. I think it may be of interest to Dr. Isaac Asimov, whose address is . . .
                                        Yours faithfully,

This usually ends the matter, as far as I am concerned, but that brave try at sabotage in the last sentence has so far failed completely. The competition is still six books ahead.

Occasionally, something a little more drastic in the way of evasive action may be necessary—as when, for example, the writer informs me that his thirty thousand word explanation of the rings of Saturn, the tides, the weather, and the rise and fall of the stock market is merely a prelude to the full five volume work, just waiting to be mailed to me.

Then I may take the coward's way out, and the writer gets this:

Dear Sir,
   This is to acknowledge your letter to Mr. Clarke. I will bring it to his notice as soon as he is again permitted to deal with his mail.
                                        Yours faithfully,
                                        Pauline de Sylva
                                        (Secretary to Mr. Clarke)

This leaves the correspondent completely up in the air, not knowing whether I am being sobered up from the D.T.'s, serving a jail sentence, or undergoing electroshock therapy. It never fails to work.

Exceptionally long and tedious letters get the classic

treatment: "Dear Sir, You may be right. Yours faithfully . . ." And I am still keeping in reserve, for a suitable occasion, a reply once made in the House of Lords: "I am indebted to you for demonstrating that there is no such thing as unutterable nonsense."

Very effective is the phony technical answer, which may run like this:

> Your treatment fails to explain the well-known fact that the locus of the contravariant tensor has non-communicative divergence in the region of the transfinite singularity. A simple extension of your theory leads at once to the obviously fallacious conclusion that the polarized proton flux will result in a heuristic phase imbalance of the hypergeometric catenery, so that . . . etc. etc.

I can go on for ages, but this sample should be quite sufficient.

The technique of fighting fire with fire is, of course, a little risky; the letter may fall into the hands of someone who will recognize it for the utter gibberish that it is. But sometimes the temptation to reply in kind is irresistible; this is how I once yielded to it, at the height of the Flying Saucer craze:

> Dear Sir,
>
> You have been completely deceived; the visitors from space who landed in your back garden and informed you that they had come from the planet Ying, 50,000 light-years away, are imposters. I have definite proof that they are actually from the planet Yang, which is only *40,000* light-years away.
>
> Yours faithfully,

My proudest invention, however, is a diabolical device that I call the random noise, or zero-information, letter. This is hard work, and I employ it only when a writer has been abusive, condescending, or has used insufficient stamps, thus dragging me out of bed before dawn to pay the postman some mysteriously computed excess.

The Random letter has to be *handwritten,* not typed, and I stumbled upon the technique quite by accident when trying to interpret some notes I'd made for a story. Though the story is lost without trace, I have something that is probably more valuable. In the zero-information letter, all the neutral words are *just* legible, but the key ones are not. In fact, they aren't even words, but only realistic squiggles which look—at least in my handwriting—as if they ought to be words. I can best demonstrate by another example:

Dear Sir,

Thank you for your squiggly letter of the 30th February, which has caused me much squiggling. It is really astonishing that you have *completely* squiggled my squiggle! In particular, your statement that squiggle *squiggle* is actually *squiggle* squiggle is certainly one of the most remarkable pieces of squiggle that I have ever squiggled! SQUIGGLE SQUIGGLE TO YOU!! I cannot too strongly suggest that you squiggle at the earliest possible moment. . . . etc., etc.

Yours squigglishly,
Arthur C. Squiggle

This is not as easy as it looks, since sense keeps squiggling—er, creeping—in if I am not careful. The aim, of course, is to leave the maddened recipient completely unable to decide whether I'm expressing enthusiastic agreement, or telling him to go boil his head. The lavish use of capitals, exclamation marks and underlining strengthens the impression that I'm saying something of the utmost importance. At the same time, I leave no hope that the meaning will ever be discovered; any further communications will obviously be even less decipherable.

Serious letters, no matter how much I disagree with

them, get a serious reply; but if castigation appears necessary, I do not hesitate. Nor do I indulge in any hypocritical "This hurts me more than it hurts you" line of talk, for one of my minor pleasures is not suffering fools gladly. As William Blake so wisely remarked, "Damn braces; Bless relaxes," and a good anathematization is excellent for the system.

The last time I enjoyed one was when a well-known national monthly published an article in which I outlined the useful and exciting things we could do when we reached the Moon. This so shook the magazine's less science-oriented readers that some of them wanted to know if it was a joke. Soon afterwards, the President of the United States announced that the Moon was the prime goal of the American space effort, so that question was answered. However, what really annoyed me was a letter from an engineer actually in the astronautics industry, who accused me of spreading technical misinformation and suggested that I should "terminate the role of popularizer, for which I was ill-suited."

He had selected an unfortunate example: the essay that had aroused his ire had just been chosen as runner-up for the Westinghouse–AAAS Award as the best magazine science article of the year.[1] While informing my critic of this, I also tactfully hinted that there wasn't much hope for the U.S. space effort if *he* was associated with it.

Some close-in knife work was also called for not long ago, when I received a booklet from a gentleman who claimed to have solved that classic problem of Greek geometry, the squaring of the circle. Now, one can construct a square approximately equal in area to a given circle, to any desired degree of accuracy. But it is impossible, using only compass and straightedge, to do so *exactly*, though it took more than two thousand years to demonstrate this. (The final proof was not obtained until 1882.) Yet, through ignorance or vanity,

---

[1] The article is "The Uses of the Moon," in this volume.

men continue on a quest which is now known to be as fruitless as the search for perpetual motion.

I would not have bothered with this particular circle squarer, but he was so proud of his achievement that he had got it into the *Congressional Record*. I will reveal neither the name, nor the state, of the hapless legislator involved in this waste of taxpayers' money, but will merely hint that he probably knew more about pineapples than geometry. Any bright high school boy could have found the error in this "proof," extolled in the *Record* as one of the mathematical triumphs of the ages.

As it seems to me (and to a good many other people) that the scientific education of Congressmen is a pressing need, I sent the carbon copy of my disproof to Washington. Though I have occasionally, to my surprise, appeared in the *Congressional Record,* I do not know if this amendment was ever inserted. I rather doubt it.

Very rarely there is a borderline letter that does not fall into any of the above categories. It may be from somebody who has got hold of a bright idea, or a useful invention, but can get no one to take any interest in it. He has grown heartily sick of the scientific and bureaucratic runaround; can I suggest anyone who will help?

It is very seldom that I can, even when I am wholly sympathetic toward the project concerned. To quote an example: Recently I received a letter from a consulting engineer who had some quite interesting ideas on the nature of telepathy, and was anxious to get them tested. I could only reply that I would be delighted to see some real progress on this front, but had long ago refused to get involved with ESP, as it seems a bottomless morass. Perhaps this was a lazy man's answer; but I have a living to earn.

I hope that this has not given the impression that I don't *like* hearing from readers. Far from it, for I must now overcome my natural modesty and admit that the great majority are straightforward letters of appreciation for the pleasure my books have given. These always receive a prompt "thank you," carefully listing any recent titles that may have been overlooked. Such letters present no problems, and are of no interest to anyone except the sender and myself. I enjoy getting them, but do not think that it would make very much difference to my output, my style of writing, or my choice of plots if there was no feedback at all from the audience. Nor have I ever been responsive to entreaties for sequels to any of my short stories or novels; I really mean it, when I write THE END.

Appendix

EXTRATERRESTRIAL RELAYS

This article is reprinted here exactly as it appeared in *Wireless World* for October 1945, and I have made no changes in the text. There has been no prior publication in book form, so it has taken just twenty years to attain the dignity of hard covers.

EXTRATERRESTRIAL RELAYS

Can Rocket Stations Give
Worldwide Radio Coverage?

Although it is possible, by a suitable choice of frequencies and routes, to provide telephony circuits between any two points or regions of the Earth for a large part of the time, long-distance communication is greatly hampered by the peculiarities of the ionosphere, and there are even occasions when it may be impossible. A true broadcast service, giving constant field strength at all times over the whole globe would be invaluable, not to say indispensable, in a world society.

Unsatisfactory though the telephony and telegraph position is, that of television is far worse, since ionospheric transmission cannot be employed at all. The service area of a television station, even on a very good site, is only about a hundred miles across. To cover a small country such as great Britain would require a network of transmitters, connected by coaxial lines, wave guides, or VHF relay links. A recent theoretical study[1] has shown that such a system would require repeaters at intervals of fifty miles or less. A system of this kind could provide television coverage, at a very considerable cost, over the whole of a small country. It would be out of the question to provide a large continent with such a service, and only the main centers of population could be included in the network.

The problem is equally serious when an attempt is made to link television services in different parts of the globe. A relay chain several thousand miles long would

230

cost millions, and transoceanic service would still be impossible. Similar considerations apply to the provision of wide-band frequency modulation and other services, such as high-speed facsimile which are by their nature restricted to the ultra-high frequencies.

Many may consider the solution proposed in this discussion too farfetched to be taken very seriously. Such an attitude is unreasonable, as everything envisaged here is a logical extension of developments in the last ten years—in particular the perfection of the long-range rocket of which V-2 was the prototype. While this article was being written, it was announced that the Germans were considering a similar project, which they believed possible within fifty to a hundred years.

Before proceeding further, it is necessary to discuss briefly certain fundamental laws of rocket propulsion and "astronautics." A rocket which achieved a sufficiently great speed in flight outside the Earth's atmosphere would never return. This "orbital" velocity is 8 km per sec. (5 miles per sec.), and a rocket which attained it would become an artificial satellite, circling the world forever with no expenditure of power—a second moon, in fact. The German transatlantic rocket A.10 would have reached more than half this velocity.

It will be possible in a few more years to build radio-controlled rockets which can be steered into such orbits beyond the limits of the atmosphere and left to broadcast scientific information back to the Earth. A little later, manned rockets will be able to make similar flights with sufficient excess power to break the orbit and return to Earth.

There are an infinite number of possible stable orbits, circular and elliptical, in which a rocket would remain if the initial conditions were correct. The velocity of 8 km/sec. applies only to the closest possible orbit, one just outside the atmosphere, and the period of revolution would be about ninety minutes. As the radius of the orbit increases, the velocity decreases, since gravity is diminishing and less centrifugal force is needed to balance it. Fig. 1 shows this graphically. The Moon, of course, is a particular

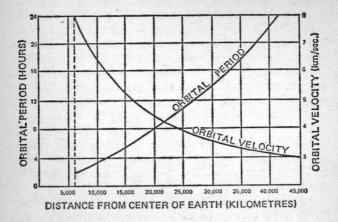

Fig. 1. Variation of orbital period and velocity with distance from the center of the earth.

case and would lie on the curves of Fig. 1 if they were produced. The proposed German space stations would have a period of about four and a half hours.

It will be observed that one orbit, with a radius of 42,000 km, has a period of exactly twenty-four hours. A body in such an orbit, if its plane coincided with that of the Earth's equator, would revolve with the Earth and would thus be stationary above the same spot on the planet. It would remain fixed in the sky of a whole hemisphere and unlike all other heavenly bodies would neither rise nor set. A body in a smaller orbit would revolve more quickly than the Earth and so would rise in the west, as indeed happens with the inner moon of Mars.

Using material ferried up by rockets, it would be possible to construct a "space station" in such an orbit. The station could be provided with living quarters, laboratories, and everything needed for the comfort of its crew, who would be relieved and provisioned by a regular rocket service. This project might be undertaken for purely scientific reasons as it would contribute enormously to our knowledge of astronomy, physics, and meteorology. A good deal of literature has already been written on the subject.[2]

Although such an undertaking may seem fantastic, it requires for its fulfillment rockets only twice as fast as those already in the design stage. Since the gravitational stresses involved in the structure are negligible, only the very lightest materials would be necessary and the station could be as large as required.

Let us now suppose that such a station were built in this orbit. It could be provided with receiving and transmitting equipment (the problem of power will be discussed later) and could act as a repeater to relay transmissions between any two points on the hemisphere beneath, using any frequency which will penetrate the ionosphere. If directive arrays were used, the power requirements would be very small, as direct line of sight transmission would be used. There is the further important point that arrays on the Earth, once set up, could remain fixed indefinitely.

Moreover, a transmission received from any point on the hemisphere could be broadcast to the whole of the visible face of the globe, and thus the requirements of all possible services would be met (Fig. 2).

It may be argued that we have as yet no direct evidence of radio waves passing between the surface of the Earth and outer space; all we can say with certainty is that the shorter wavelengths are not reflected back to the Earth. Direct evidence of field strength above the Earth's atmosphere could be obtained by V-2 rocket technique, and it is to be hoped that someone will do something about this soon as there must be quite a surplus stock somewhere! Alternatively, given sufficient transmitting power, we might obtain the necessary evidence by exploring for echoes from the Moon. In the meantime we have visual evidence that frequencies at the optical end of the spectrum pass through with little absorption except at certain frequencies at which resonance effects occur. Medium high frequencies go through the E layer twice to be reflected from the F layer and echoes have been received from meteors in or above the F layer. It seems fairly certain

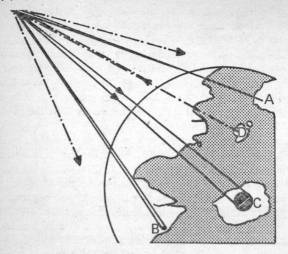

Fig. 2. Typical extraterrestrial relay services. Transmission from A being relayed to point B and area C; transmission from D being relayed to whole hemisphere.

that frequencies from, say, 50 Mc/s to 100,000 Mc/s could be used without undue absorption in the atmosphere or the ionosphere.

A single station could only provide coverage to half the globe, and for a world service three would be required, though more could be readily utilized. Fig. 3 shows the simplest arrangement. The stations would be arranged approximately equidistant around the Earth, and the following longitudes appear to be suitable:

The stations in the chain would be linked by radio or optical beams, and thus any conceivable beam or broadcast service could be provided.

The technical problems involved in the design of such stations are extremely interesting,[3] but only a few can be gone into here. Batteries of parabolic reflectors would be provided, of apertures depending on the frequencies employed. Assuming the use of 1,000 Mc/s waves, mirrors about a meter across would beam almost all the power on to the Earth. Larger reflectors could be used to illumi-

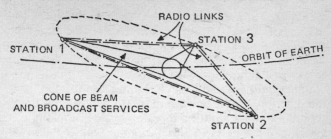

Fig. 3. Three satellite stations would insure complete coverage of the globe.

> 30 E—Africa and Europe
> 150 E—China and Oceana
> 90 W—The Americas

nate single countries or regions for the more restricted services, with consequent economy of power. On the higher frequencies it is not difficult to produce beams less than a degree in width, and, as mentioned before, there would be no physical limitations on the size of the mirrors. (From the space station, the disk of the Earth would be a little over 17 degrees across.) The same mirrors could be used for many different transmissions if precautions were taken to avoid cross modulation.

It is clear from the nature of the system that the power needed will be much less than that required for any other arrangement, since all the energy radiated can be uniformly distributed over the service area, and none is wasted. An approximate estimate of the power required for the broadcast service from a single station can be made as follows:

The field strength in the equatorial plane of a $\lambda/2$ dipole in free space at a distance of d meters is $e = 6.85 \dfrac{\sqrt{P}}{d}$ volts/meter where P is the power radiated in watts.

Taking d as 42,000 km (effectively it would be less), we have $P = 37.6e^2$ watts. (e now in μV/meter.)

If we assume e to be 50 microvolts/meter, which is the F.C.C. standard for frequency modulation,[4] P will be

94 kW. This is the power required for a single dipole, and not an array which would concentrate all the power on the Earth. Such an array would have a gain over a simple dipole of about 80. The power required for the broadcast service would thus be about 1.2 kW.

Ridiculously small though it is, this figure is probably much too generous. Small parabolas about a foot in diameter would be used for receiving at the Earth end and would give a very good signal/noise ration. There would be very little interference, partly because of the frequency used and partly because the mirrors would be pointing toward the sky which could contain no other source of signal. A field strength of 10 microvolts/meter might well be ample, and this would require a transmitter output of only 50 watts.

When it is remembered that these figures relate to the broadcast service, the efficiency of the system will be realized. The point-to-point beam transmissions might need powers of only 10 watts or so. These figures, of course, would need correction for ionospheric and atmospheric absorption, but that would be quite small over most of the band. The slight falling off in field strength due to this cause toward the edge of the service area could be readily corrected by a nonuniform radiator.

The efficiency of the system is strikingly revealed when we consider that the London television service required about 3 kW average power for an area less than fifty miles in radius.[5]

A second fundamental problem is the provision of electrical energy to run the large number of transmitters required for the different services. In space beyond the atmosphere, a square meter normal to the solar radiation intercepts 1.35 kW of energy.[6] Solar engines have already been devised for terrestrial use and are an economic proposition in tropical countries. They employ mirrors to concentrate sunlight on the boiler of a low-pressure steam engine. Although this arrangement is not very efficient it could be made much more so in space where the operating components are in a vacuum, the radiation is intense and continuous, and the low-temperature end of the cycle could be not far from absolute zero. Thermoelectric and photoelectric developments may make it possible to utilize the solar energy more directly.

Though there is no limit to the size of the mirrors that could be built, one fifty meters in radius would intercept over 10,000 kW and at least a quarter of this energy should be available for use.

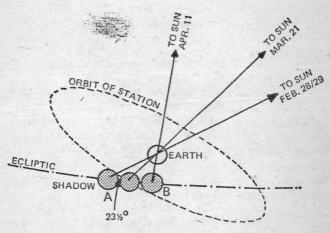

Fig. 4. Solar radiation would be cut off for a short period each day at the equinoxes.

The station would be in continuous sunlight except for some weeks around the equinoxes, when it would enter the Earth's shadow for a few minutes every day. Fig. 4 shows the state of affairs during the eclipse period. For this calculation, it is legitimate to consider the Earth as fixed and the Sun as moving round it. The station would graze the Earth's shadow at A, on the last day in February. Every day, as it made its diurnal revolution, it would cut more deeply into the shadow, undergoing its period of maximum eclipse on March 21st. On that day it would only be in darkness for one hour, nine minutes. From then onward the period of eclipse would shorten, and after April 11th (B) the station would be in continuous sunlight again until the same thing happened six months later at the autumn equinox, between September 12th and October 14th. The total period of darkness would be about two days per year, and as the longest period of eclipse would be little more than an hour there should be no difficulty in storing enough power for an uninterrupted service.

CONCLUSION

Briefly summarized, the advantages of the space station are as follows:

1. It is the only way in which true world coverage can be achieved for all possible types of service.

2. It permits unrestricted use of a band at least 100,000 Mc/s wide, and with the use of beams an almost unlimited number of channels would be available.

3. The power requirements are extremely small since the efficiency of "illumination" will be almost 100 per cent. Moreover, the cost of the power would be very low.

4. However great the initial expense, it would only be a fraction of that required for all the world networks replaced, and the running costs would be incomparably less.

APPENDIX—ROCKET DESIGN

The development of rockets sufficiently powerful to reach "orbital" and even "escape" velocity is now only a matter of years. The following figures may be of interest in this connection.

The rocket has to acquire a final velocity of 8 km/sec. Allowing 2 km/sec. for navigational corrections and air resistance loss (this is legitimate as all space rockets will be launched from very high country) gives a total velocity needed of 10 km/sec. The fundamental equation of rocket motion is

$$V = v \log_e R$$

where V is the final velocity of the rocket, *v* the exhaust velocity and R the ratio of initial mass to final mass (payload plus structure). So far *v* has been about 2–2.5 km/sec. for liquid fuel rockets but new designs and fuels will permit of considerably higher figures. (Oxyhydrogen fuel has a theoretical exhaust velocity of 5.2 km/sec. and more powerful combinations are known.) If we assume *v* to be 3.3 km/sec., R will be 20 to 1. However, owing to its finite acceleration, the rocket loses velocity as a result of gravitational retardation. If its acceleration (assumed constant) is *a* meters/sec², then the necessary ratio R_g is increased to

$$R_g = (R)^{\frac{a+g}{a}}$$

For an automatically controlled rocket *a* would be about 5g and so the necessary R would be 37 to 1. Such ratios cannot be realized with a single rocket but can be attained by "step rockets"[2] while very much higher ratios (up to 1,000 to 1) can be achieved by the principle of "cellular construction."[7]

EPILOGUE—ATOMIC POWER

The advent of atomic power has at one bound brought space travel half a century nearer. It seems unlikely that we will have to wait as much as twenty years before atomic-powered rockets are developed, and such rockets could reach even the remoter planets with a fantastically small fuel/mass ratio—only a few per cent. The equations developed in the appendix still hold, but *v* will be increased by a factor of about a thousand.

In view of these facts, it appears hardly worth while to expend much effort on the building of long-distance relay chains. Even the local networks which will soon be under construction may have a working life of only twenty to thirty years.

REFERENCES

1. "Radio-Relay Systems," C. W. Hansell. Proc. I.R.E., Vol. 33, March 1945.
2. "Rockets," Willy Ley. (Viking Press, N.Y.)
3. "Das Problem der Befahrung des Weltraums," Hermann Noordung.
4. "Frequency Modulation," A. Hund. (McGraw-Hill.)
5. "London Television Service," MacNamara and Birkinshaw. J.I.E.E., Dec. 1938.
6. "The Sun," C. G. Abbot. (Appleton-Century Co.)
7. *Journal of the British Interplanetary Society*, Jan. 1939.